P9-DDV-532

Hormonal

Hormonal

THE HIDDEN INTELLIGENCE OF HORMONES

how they drive desire, shape relationships, influence our choices, and make us wiser

MARTIE HASELTON, PhD

LITTLE, BROWN AND COMPANY
New York • Boston • London

Little, Brown and Company
Hachette Book Group
1290 Avenue of the Americas, New York, NY 10104
littlebrown.com

First Edition: February 2018

Little, Brown and Company is a division of Hachette Book Group, Inc. The Little, Brown name and logo are trademarks of Hachette Book Group, Inc.

Images on page 70 reprinted from Jones et al., "Menstrual Cycle, Pregnancy and Oral Contraceptive Use Alter Attraction to Apparent Health in Faces," 2005, Proceedings of the Royal Society of London B: *Biological Sciences*, vol. 272, issue 1561, 347–54, by permission of the Royal Society; page 75, reprinted from MJ Law Smith et al., "Facial Appearance Is a Cue to Oestrogen Levels in Women," 2006, Proceedings of the Royal Society of London B: *Biological Sciences*, vol. 272, issue 1583, 135–40, by permission of the Royal Society; page 118 (top), reprinted from *Hormones and Behavior*, vol. 90, James Roney and Zachary Simmons, "Ovarian Hormone Fluctuation Predict Within-Cycle Shifts in Women's Food Intake," 8–14, copyright 2017 with permission from Elsevier; page 118 (bottom), reprinted from *Hormones and Behavior*, vol. 6, issue 4, John Czaja and Robert W. Goy, "Ovarian Hormones and Food Intake in Female Guinea Pigs and Rhesus Monkeys," 21, copyright 1975 with permission from Elsevier; page 127, from Kristina Durante, Norman P. Li, and Martie G. Haselton, "Changes in Women's Choice of Dress Across the Ovulatory Cycle: Naturalistic and Laboratory Task-Based Evidence," *Personality and Social Psychology* (volume 41, issue 11), pages 1451–60. Copyright © 2008. Reprinted by permission of SAGE publications; page 141, from Debra Lieberman, Elizabeth G. Pillsworth, and Martie G. Haselton, "Kin Affiliation Across the Ovulatory Cycle: Females Avoid Fathers When Fertile," *Psychological Science* (volume 22, issue 1). Copyright © 2011. Reprinted by permission of SAGE publications; page 209, reprinted from *Trends in Ecology and Evolution*, volume 25, issue 3, Alexandra Alvergne and Virpi Lummaa, "Does the Contraceptive Pill Alter Mate Choice in Humans?," 171–79, copyright 2010 with permission from Elsevier.

ISBN 978-0-316-36921-3
LCCN 2017953894

10 9 8 7 6 5 4 3 2 1

LSC-C

Printed in the United States of America

To my father, Mark Barden Haselton. I miss you.

*And to my family, Jackie and Rick Seib, Pamela Haselton,
Jodie, Tim, Taylor, and Billie Niznik, and most of all to my children,
Georgia and Lachlan*

Contents

Hormonal

Introduction

The New Darwinian Feminism

I HAVE HAD THE great fortune of finding a scientist's dream: a fascinating, socially relevant topic that was underexplored but rich in the "empirical goods" it delivered, once I started doing research.

At the time I began my work on women's hormone cycles, the broad consensus in science was that humans were vastly different from other species and that hormone cycles didn't have much to do with patterns of human sexual behavior. Everyone thought that humans had been "emancipated" from hormonal control, whereas our nonhuman cousins were still seized by it. Part of that thinking was based on a genuine recognition that humans had unusual attributes, including this one: Humans will have sex at nearly any time — when at peak fertility near ovulation, but also at other times in the cycle, and even when conception is impossible, for example, when a woman is pregnant, newly postpartum and breastfeeding, and in the postreproductive years following menopause. This sort of "extended sexuality" runs in stark contrast to patterns in other mammals.

I certainly recognized that women were different. We're not automatons who experience an estrogen surge and then follow a behavioral urge — as in having sex (or competing with a rival).

But my training as an evolutionary theorist led me to consider that hormones might affect women too, guiding their sexual and social decisions. Hormones control reproduction, the powerful engine of natural selection. It therefore seemed astonishingly unlikely that hormones would not guide our behavior in some way.

The first studies from my lab found that women at peak fertility seemed to place a premium on sexiness in male partners. They also felt more attractive and wanted to go out to clubs and parties where they might meet men. They even showed up at the lab wearing dressier and sometimes more revealing clothing. It appeared that ovulation might be triggering "mate shopping."

Although I initially viewed this research as a side project, these findings were too tantalizing in their implications for me to pass up — I had to pursue them further. What other under-the-radar secrets of female desire might I discover if I kept looking? And so I did, with the help of dozens of students and many cherished colleagues.

I am writing this book to share the story of our fascinating findings. It turns out that we are hormonally intelligent. Our hormones affect everything from our mating desires (Chapters 2 and 4) to our competitive urges (Chapter 5), our physical and behavioral changes during pregnancy and new motherhood (Chapter 7), and our "next chapter" — menopause — with its potential for liberating new experiences beyond reproduction (Chapter 7).

I am also writing this book as a call to action for more information on our female brains and bodies. In spite of all that we do know, research was stunted for decades by the notion that men could be considered the "default sex" for biomedical information (*If it worked for men, why wouldn't it work for women too?*). Women, the thinking went, were just too "messy" with all those hormone cycles. Why bother?

I believe we know far too little about women's hormones and

their behavior, and we must know more so that we can make the best decisions at every phase of life. What are the consequences of suppressing our hormone cycles — or even eliminating menstrual periods altogether — by using the pill? What's the truth about getting pregnant as we move into our third, fourth, and even fifth decades? Can we find a magic bullet for female desire, the way Viagra helped men with their bedroom troubles? Should we consider taking hormone supplements later in life? I address all these questions in the book, but we still do not have complete answers, to the detriment of both women and men.

I also want to offer a different perspective on women and hormones. Outside of academic circles, for decades women have been derided for being "hormonal" — even too "hormonal" to be president. (For the rest of my days, I will never forget bringing my young daughter with me as I cast my vote on November 8, 2016, confident that she was witnessing history with the election of the first female president — *ouch.*) One would think these chauvinistic views are antiquated, but I have seen them come to the fore again and again, as I detail in Chapter 1.

There's a double standard in our views on hormone cycles in men and women. Gloria Steinem highlighted it in an essay from the late eighties called "If Men Could Menstruate."[1] She argued that if men were the ones who got menstrual periods, periods would become a source of masculine pride. She also said, "Sanitary supplies would be federally funded and free." And, hilariously, "Of course, some men would still pay for the prestige of such commercial brands as Paul Newman Tampons, Muhammad Ali's Rope-a-Dope Pads, John Wayne Maxi Pads, and Joe Namath Jock Shields — 'For Those Light Bachelor Days.'" (For more on the movement to reduce the "pink tax," see "Free the Tampons!" in Chapter 3.)

Some believe that any biological explanation for a woman's

behavior will keep her from achieving. The thinking is that differences between women and men, if they have any hint of a biological foundation, will doom women to girlish stereotypes, confine them to a maternal role, and smash them up against the glass ceiling if they try to achieve professionally. The implication for researchers is that we should keep information about women's hormones and their behavior on the down low. It's best not to stir up these stereotypes.

I think the opposite is true. We do not help women by suppressing information or failing to do the research that could inform the important hormonal questions we need answers to. And what we've learned about women and hormones is — in my view — empowering. It's not a simple story about going "hormonal" on those last few days of the cycle and losing our rational faculties. It's the story of how our hormones guide us through uniquely female life experiences — from feeling desire and pleasure to choosing a mate, having a child (if we would like to), raising a child, and transitioning to our postreproductive years. These experiences are crucial to our understanding of what it means to be human. They also unite us with our mammalian cousins and even the colossal lizards that once roamed the earth. To be sure, we do things in our own exclusively human way, and I will explain why in Chapter 4.

When I started my career in science I thought I could never allow politics to creep into my work. I aspired to be the objective scientist. Just the facts, please! But politics, or at least controversy, seemed to follow me wherever I went. The application of evolutionary thinking to human psychology was (and is) controversial. Any biological explanation for behavior is welcomed with unease in social science, a fact that has come back to bite me... many times! But my results seemed to provide powerful evidence of the footprints of evolution on the human social mind. So the findings were newsworthy not only because they were new and sexy but also because of their deeper implications for understanding the forces that have shaped our minds and behaviors. The work helped to put me on the map as a scientist.

I've fought my battles. I would publish research findings after doing careful work, taking the long road to try to get the best data rather than the quick and easy route. Pushback would come in the form of published critiques of the scientific methods we used in my research lab and insinuations that the results were really just due to iffy statistical practices that produce the illusion of findings. (In fact, when we looked at the evidence pertinent to those claims, it flatly contradicted what the critics were saying. I am still waiting for the e-mail that says, *Oops, our bad!*) I have been shouted down at conferences and have gotten some jaw-dropping e-mails. I will not say that I have a lock on all of the scientific facts — I have a healthy dose of skepticism about the data from my lab and the work of others. But I will say that some of the conflict was way over the top — seemingly worthy of a Shakespearean drama (had the Bard set one of his plays at a large American research university or two).

My brushes with controversy started early in my career. As an undergraduate, I knew I wanted to be a psychologist, but I was also keenly interested in what I viewed as more "hard-core," biology-based scientific explanations for human behavior, which were not part of the mainstream at the time. I had an enlightening experience in a philosophy course that forecast my scientific path.

The professor explained the difference between dualism (separate explanations for "mind" and "body," with some sort of gremlin driving the machine of the mind) and materialism (the brain produces behavior, period!). He asked for a show of hands.

Who is a dualist? Every hand in the room went up — except mine. *Who is a materialist?* I enthusiastically raised my hand and looked around at my classmates, who now seemed like utter buffoons to me. From that point on, I was on a mission to detect bullshit and squash it.

In my early graduate school years, I had an encounter with the famous evolutionary biologist Stephen Jay Gould, who was known for many things, including drawing a line between evolutionary explanations for the grand variety of physical forms of life and explanations for human behavior. I had just started to study evolutionary psychology (which I discuss in Chapter 2) and found its logic so compelling. Yes, there were evolutionary explanations for our physical bodies and organ systems — but in this field I was studying there were evolutionary explanations for our mental "organs" (and resulting behaviors) as well!

I snuck into the Q&A Gould was giving to biology students. I raised my hand and asked him why he thought evolutionary psychology was problematic. This was clearly not a question he was expecting to get, and he uncharacteristically meandered a little bit with his answer, talking about ideas being difficult to test and so on. Despite the fact that I was extremely nervous to speak up in front of hundreds of biology graduate students — let alone in front of Gould, with his formidable persona — I pressed him. I asked, well, then, why do we see such robust patterns across thirty-seven cultures around the globe in differences between what men want and what women want in mates? (I discuss this in Chapter 4.) He said, hmm, well, maybe there is something to that. Score one for the bullshit squasher! I was hooked.

I tell these stories not because I object to political activism, and certainly not because I object to the sentiments that drive people to want to tear down barriers to equal opportunity. I share these sentiments because I am a feminist. I believe women do not have all of the opportunities that men have, particularly in business, government, and science. (I also believe that women have some important opportunities that men do not have, though perhaps they are fewer in number.) I recognize that there are stereotypes, even among enlightened feminists of both sexes, that cause us to unfairly judge men and women. But I want to argue for a new breed of feminism, a new *Darwinian* feminism.

This kind of feminism respects our biology and fully explores it. Women have a right to understand the history — including the evolutionary history — that has shaped our bodies and minds. We need better information about our biological — and hormonal — natures. Yes, some people will take a simplistic view or perhaps be motivated by sexism to claim that women's biology "is destiny." But if we have learned anything, it is that although biology plays a role, our social context (and our agency to reflect and make choices) matters just as much. So, I think we stand up to people who make those simplistic arguments. We set them straight. We say, yes, there's a biological foundation — for the behavior of both women *and* men. But it's best to understand it rather than to be ignorant, don't you think?

I also believe that the more we understand the origins and workings of our hormonal nudges, the better we are able to channel them (or ignore them, if we so choose). That's a key message of this book (and particularly of Chapters 7 and 8).

I have another motivation for writing this book: my students, and in particular the students who have been seated in my classroom as first-year undergraduates taking the so-called sex cluster, an interdisciplinary course on sex and gender that I teach with wonderful co-instructors at the University of California, Los Angeles. The course attracts a lot of female students as well as a lot of gender-atypical students who are drawn to the course for the kind of research we call "*me*-search." I have wanted to show them that they can do science if they want to, even if they don't look or act like the stock photo of the (usually male) scientist in the white lab coat. At the end of the academic term we hold a roundtable with all of the professors and graduate-student instructors. We voice our opinions and talk about pertinent research, but first we ask students what they think. We ask questions like, "Should we enforce a gender balance so that there are as many women as men enrolled in physics?" (The students' answer is almost always no. They want choice — not to be forced.)

For the past few years as part of this discussion I've told a story. I had just begun graduate school, and I recognized that being feminine conflicted with that stock photo image in ways that could undermine my own scientific credibility. So, I made a conscious choice to tone it down — "it" being my own feminine appearance. No makeup. Jeans and sweater and sneakers. Droopy hair straight out of the shower. I wanted to be taken seriously, to gain acceptance. But after a while, I felt like I was wearing a bad disguise, one that seemed disingenuous. Then, as I tell them, "One day I said, *fuck it!* I am going to be who I am and look how I look, and if I have to work harder because I am a woman doing science, fine." I refused to be ensnared in a misguided female stereotype.

As I hope this book will show, no woman should be held back by a misguided hormonal stereotype. In fact, I think we should take back the word "hormonal" — after all, we *are* — and celebrate it, because our hormones have the potential to give us pleasure, guide us through life, and make us all the wiser.

1

The Trouble with Hormones

"DON'T ASK HOLLY FOR a raise today; she'll eat you alive." Why's that? *She's hormonal.* "She's happy with everything one minute, but then she gets so upset the next." *She's hormonal.* "Whoa, she's coming on to everything with a pulse." *She's hormonal.*

Difficult female... baby machine... crazy lady... red-hot momma... ice queen... bitch on wheels.

No matter how modern and progressive an age we may live in (or think we live in), all those female stereotypes are still very much alive today. They most certainly didn't vanish in the last century as women entered the workforce in record numbers, rising to leadership positions in virtually every field and eventually overtaking the number of males who graduate from American colleges and universities.[1]

But those aren't just ordinary stereotypes, because they have a biological component that sets them apart from the other tired notions like *damsel in distress.* Those perceptions of women — as well as many others that crop up in everyday situations at work, at home, at school — spring from one central idea: that female hormones control female behavior, as in *She's hormonal.*

She's hormonal — her monthly fluctuating levels of estrogen and other female reproductive hormones are making her act this way. But that's not a very catchy thing to say.

Here's the truth: Perhaps she *is* hormonal, but all humans, male and female, have hormonal cycles. (No one would ever say that a man is hormonal, at least not with the same negative connotations, even though testosterone levels have a daily up-and-down cycle, not a monthly one.) It might be only a tad more decorous than "she's on the rag," but women seem to own the "hormonal" label.

Here's the problem: Explaining female behavior — particularly behavior that is seen as overly aggressive, unbalanced, or somehow out of character for that girl or woman — by attributing it to sex hormones is a gross and damaging oversimplification. In essence, it says that females have little to no control over their actions because they are governed by their biology. But that dumbed-down interpretation overshadows something valuable, important, and life changing for both women and men.

In fact, women's hormone cycles embody half a billion years of evolutionary wisdom. While hormones most definitely influence female behavior — I am writing a whole book about it, after all — there is a hidden intelligence embedded in the female fertility cycle: an ancient knowledge that women can use to make the best decisions in their modern lives. Behind the everyday behavior that some interpret as simply "hormonal," there is a biochemical process that has helped females — billions across thousands of species — choose mates, avoid rape, compete with female rivals, fight for resources, and produce offspring with fit genes and good prospects. To master these challenges, female brains evolved to conspire with their hormones rather than be corrupted by them.

Hormones are a crucial reason we've survived and thrived.

Biology Is Not Destiny (but It Is Political)

As a scientist and as a feminist, I've learned that any discussion of female hormones and the role they play in a woman's behavior can be rough terrain to negotiate, even in circles of like-minded thinkers. At first this took me by surprise — I thought everyone would want to have the benefit of this knowledge, especially women. We're entitled to understand how our bodies and minds work and why. But I've come to see that the facts get cherry-picked, then lost in a volatile mix of sexual politics. Misinformed sexists still find a way to twist the truth and use biological differences as a hurdle too high for women to clear. Feminists rightly don't want that to happen. And because of this dynamic, it becomes difficult to untangle myth from reality.

Take, for example, a highly controversial CNN story from the 2012 election year, when hormones themselves seemingly went to the polls. Two weeks before the election, the network published a story on its website reporting that according to a soon-to-be-released study,[2] during ovulation (when fertility is highest), single women favored President Barack Obama and his policies over those of Governor Mitt Romney. The story explained the researchers' findings this way: "When women are ovulating, they 'feel sexier,' and therefore lean more toward liberal attitudes on abortion and marriage equality."[3] Meanwhile, married women or those in committed relationships tilted toward the more conservative Romney, said the story.

The backlash was fast and furious, thanks to rapid-fire blogs and Internet news. "CNN thinks crazy ladies vote with their vaginas," wrote Jezebel's Katie Baker. "Hot for Obama, but Only When This Smug Married Is Not Ovulating," was the headline for Kate Clancy's response on *Scientific American*'s website. "This is exactly the nightmare image of Women Rampaging Through the Polls Judging Candidates by the Strength of Chin that has so bedeviled female candidates for so long," wrote the *Washington Post*'s Alexandra Petri. CNN took down the story within a few days, the reporter was mocked, and the lead author of the study was overwhelmed with hate mail.

The CNN retraction was seen as a victory for women, a mark of their political progress and a world away from 1970s politics, when a prominent political figure declared that women's "raging hormonal influences" disqualified them from leadership positions. Dr. Edgar Berman was a member of the Democratic National Party's Committee on National Priorities and a top adviser to Vice President Hubert Humphrey, as well as his personal physician. In 1970, when a female member of Congress suggested that women's rights be given top priority within the party, Berman's negative response was dismissive and downright Victorian. He cited the menstrual cycle and menopause as reasons why women would never achieve equality.

"If you had an investment in a bank," he explained, "you wouldn't want the president of your bank making a loan under these raging hormonal influences at that particular period." He went on. "Suppose we had a President in the White House, a menopausal woman president who had to make the decision of the Bay of Pigs, which of course was a bad one, or the Russian contretemps with Cuba at the time?" Berman essentially implied that a moody female leader of the free world would pick up the red telephone in the Oval Office and bitch out the Kremlin, triggering a nuclear holocaust. (It's after the 2016 presidential election as I write this — oh, the irony.)

There is some evidence that Berman, a staunch Democrat

who advocated for women's issues like daycare and easier access to birth control, was trying to move the discussion along to issues like Vietnam and was making an attempt at humor, but if so, his listening skills were way off and his comedic timing was truly lousy. This was a crucial moment for the women's movement, when its leaders were trying to call attention to issues like equal pay and create support for the Equal Rights Amendment. Along comes Berman, who delivers a punch line that feeds right into the sexist notions of a woman's place being in the home. *The Mary Tyler Moore Show*, featuring career-minded Mary Richards, debuted in 1970, the same year Berman declared women too hormonal. But the beauty queens on *The Miss America Pageant* were still pulling in more viewers than Mary, and Samantha of *Bewitched*, for all her pluckiness, spent a lot of time cleaning the house (like a mortal).

Even without the Internet, it didn't take long for Berman's comments to go public, and within a few months he resigned from his committee position. Not only were his politics questioned, but so was his scientific knowledge. "To talk about 'raging hormonal influences' is nonsense, or at least a gross overstatement," said Harvard endocrinologist Dr. Sidney Ingbar, voicing a professional opinion that would be echoed by others. "Anyone who speaks with authority," added psychiatrist Dr. Leon J. Epstein of the University of California, "stands on a firm foundation of prejudice or unsupported conviction."[4]

Unlike CNN, however, Berman did not retract his statements and in fact he dug in his heels. Defending his comments later on, he wrote, "No physician (and most women) could possibly deny that during certain periods in the life span of many women there is stress and emotional disturbance over and beyond that occasioned by the average male. I stated that all things being equal during these periods of strain, I personally valued the male judgment in crucial decisions. . . . I cannot and shall not retract a scientific truth."[5]

Of course, there is no "scientific truth" behind Dr. Berman's comments. But he was voicing a commonly held belief about women that had persisted for generations, even centuries — that female hormones were messy and problematic, something that needed to be "fixed." Menstruation and menopause were considered embarrassing topics, and women weren't being given much information about their bodies by the medical community. There was "the curse," there was "the change," and in between there were the shadowy topics of sex, pregnancy, and childbirth.

But right around the same time that Berman made these outrageous claims, a group of women had gathered in Boston. They had just published a 193-page stapled booklet on women's reproductive health: graphic talk on sexuality, pregnancy and childbirth, abortion, and other then-taboo topics. Now they were revising their humble but daring newsprint publication into what would become the first edition of *Our Bodies, Ourselves*, the blockbuster women's health bible that changed the landscape by putting self-knowledge about the female body — and the power that came with it — directly into the hands of women.[6]

We've come a long way, baby. Still, it's worth remembering how much further we need to travel. Don't forget what then presidential candidate Donald Trump said in 2015 when he complained about a female journalist who pressed him hard on his derogatory descriptions of women; he implied that she did so because she had "blood coming out of her wherever."

In other words, forty-five years later, she's still hormonal.

The Vicious Cycle (of Hormones)

It's not just male bias that has stopped us from diving right into the good stuff about hormones and female behavior. Sometimes, it's women themselves who throw up the biggest barriers — including women who have dedicated themselves to achieving equality between men and women in the workplace.

As soon as we acknowledge a difference between the sexes, the thinking goes, we lose ground in the battle and will never be treated like equals. Instead we'll be seen as weak, vulnerable, incapable. A pregnant woman will be viewed as hearing the hormone-triggered call of motherhood and won't want to come back to work — don't bother promoting her. An older woman in menopause won't be able to give 100 percent because her sleepless, hot flash–filled nights and forgetful female brain will impact her work performance, plus she'll be really difficult to deal with some days. Don't bother promoting her, either.

An artist I know recently ran right up against this line of thinking, but in a rather surprising way. Though her work was designed to empower women, she suddenly found herself being put in her place — by other feminists. When asked by friends at a dinner party what she was working on, she described her latest project: an online art installation called *The Invisible Month*,[7] a visually whimsical presentation of how estrogen and progesterone levels affect female behavior, organized around the twenty-eight-day hormonal cycle and using a flower metaphor — budding, flowering, and wilting.

For instance, if a visitor clicked on an icon for "The Budding Phase," she would see general information about estrogen and progesterone levels as well as statements (with citations from scientific literature) like this: "Estrogen creeps higher this first week and a sense of well-being increases. Mood is elevated, sleep is level. Women experience clarity of thinking, superb ability to concentrate." If a visitor clicked later in the cycle, say, "The Flowering Phase," she might see something like this: "Women are more receptive to men now. During the ovulatory part of their cycle, women were more likely to give their phone number to random male strangers who approached them in a park." (Though the reality here is that he would have to be a very handsome stranger.) Or, in "The Wilting Phase": "Menstrual migraines reduce work productivity."

"What you're saying," said one friend with alarm, "is that women don't have free will." Added another, "That's right! What if your project gets into the wrong hands of, say, the Goldman Sachs CEO, who is persuaded that women can't be in leadership roles?" Another person added, "You're basically saying that if a woman says no, just ask her in two weeks to get a yes." And the criticism kept coming. She was stunned at their response. These were all successful, educated, and forward-thinking individuals, but they seemed to be telling her to shut it down, to keep this information under wraps so that she wouldn't set women back.

Her initial mission had been simple. "I created the work as a public service work," she explained, as a way to help women — and men — see how an internal process has an external impact.[8] As an artist, she says, that concept of internal/external has long captured her imagination. With *The Invisible Month,* she was creating art, and at the same time she was sharing information that she thought other women would want to have and would embrace. But to her friends, she was opening a Pandora's box of sexual politics.

I know the feeling.

Tossing Out the Biological Baby with the Sexist Bathwater

When I first began researching the hormonal cycle, my field of social science was firmly grounded in the idea that we humans, with our big, handsome brains and opposable thumbs, were mightily different from our friends in the animal kingdom. Though of course we'd acknowledged some vital evolutionary links, we drew the line when it came to our minds, our desires, and our sexual behavior. *We* had sex when we wanted to — not when nature told us to. The other mammals were at the mercy of their hormones, all in the name of reproduction. Squirrels ran around and acted insane, then jumped on each other. We exchanged phone numbers.

You could say that my research is about seeing the animal in the human, something I've done my entire career. Making that animal-human connection is not always popular in certain academic circles or within some branches of science, where the thinking is that we humans are culturally advanced beings, with complex intellects and emotions interwoven with free will. For hundreds of years, countless scientists have labored to establish the very real differences between the human and the animal. Entire fields of study, and society itself, are built around the concept that human nature is what makes us special, different, better. If we do something, we do it for our own reasons — not because we're at the mercy of some chemical reaction triggered by sex hormones. We operate out of elegant human nature, not base animal instincts.

But the discoveries from my lab suggest that fertile women seek out the most attractive men — just as it happens among female and male primates, hamsters, and many species in between. (I discuss this research in depth in Chapter 5.) I'd studied hormones and relationships in animals, noting a pattern of behavior across the species that was impossible to ignore: Very simply put, when females — monkeys, rats, cats, dogs, and others — are most likely to conceive during their hormonal cycles, they consistently behave in ways that seem designed to attract males who offer the promise of especially fit offspring — "fit" meaning that they would have been better able to survive or reproduce in ancestral environments. Obviously the manifestation varies from species to species, but I couldn't accept that humans were entirely exempt from this predictable physical phenomenon.

Women are able to conceive for only a few days a month, making human fertility somewhat fragile and fleeting. Why *wouldn't* we have a way of making the best sexual decisions at this crucial time? Beginning in 2006, I started publishing research to show that in fact women do alter their behavior during "peak fertility." Among my findings: Women's motivation to go to clubs and parties went up, they began to notice men *other than* their "primary mates," their voices rose to a higher and more feminine pitch, they dressed in more attractive clothing, and their body odors were more attractive to men.[9] I was going right up against the assumption that human sexual behavior had been "emancipated" from hormonal control. Instead, I was suggesting that human female behavior during fertility echoes animal behavior — women's sexual desires changed and there were outward indications of women's fertility. So, fertility was not entirely concealed but on display, albeit in a more limited way than in our primate cousins.

I quickly realized that some humans don't like to be reminded that we once had tails. My research was seen as radical. To some, it was as if I was dismissing generations of scientific inquiry and saying, *Guess what, we're just a bunch of animals after all.*

Radical — and retro. Some viewed my findings as a step backward for women. They jumped on the parts of the research that made headlines because of their popular appeal — *Good Morning America* ran the headline "Does Your Cycle Make You Sexier?" — and largely ignored the broader implications of the work, which is that women brilliantly evolved their sexual behavior in response to challenges of reproduction, child-rearing, and even their own survival. But as soon as I pointed out the difference between male and female behavior, I was viewed as a water carrier for sexists. Not unlike my artist friend (or the researcher whose work blew up on CNN), I was accused of putting forth ideas that would contribute to the marginalization of women; I was calling them "hormonal."

Forty years earlier — around the time Dr. Edgar Berman published a book with one chapter titled, "The Brain That's Lame Lies Mainly in the Dame" — women were battling hard to establish feminism as a mainstream movement and bridge the gender gap by downplaying male-female differences, if not fully denying them. Starting in the late 1960s, there was a series of feminist-inspired scientific papers questioning whether PMS even existed. (And if you really want to make a woman mad, tell her that her physical and emotional discomfort is just a figment of her imagination.) For decades and in the name of equal rights, it has been considered bad form to highlight behavioral differences between men and women. It's one thing to crack a joke about men not asking for directions, but I was talking about how sex hormones affected a woman's brain.

I had offended two camps with my research: those who rejected the connection I was making between animal and human behavior, and those who rejected the line I was drawing between women and men. My research and my methods came under intense scrutiny, and detractors have even suggested that my lab massaged data, an outrageous claim.

I'd like to think the controversy has disappeared, but it hasn't, and I'm doubtful that it ever will. News producers, editors, and others who distill pop culture like to make science sexy so that people will pay attention, and as long as they're describing my work and that of others in my field with headlines like this, courtesy of the *New York Post*, it will continue to invite controversy: "Horny of Plenty: Passionate Evolutionary Biology of Human Attraction."

What probably won't make the headlines, though, is the real news here. Women's rights are enhanced — not diminished — by an increased understanding of how our bodies and minds work, and we still have much to learn. That's a large part of what motivates me. We need to know more about the impact of hormones on our relationships — including sexual and romantic relationships and our relationships with friends and kin. These are relationships that shape the entire human experience for women and men alike. We also need to better understand how hormones affect our health and feelings of well-being. But in order to learn more, we need to get more females into the lab — and not just as research scientists.

Why Viagra Was Invented for Men

For decades, important biomedical studies on diseases like cancer and drug efficacy were done on male research participants, while females were largely excluded. Even studies on stroke, more common and fatal in women than in men, once focused almost exclusively on males, and doctors didn't have enough knowledge about diagnosing heart disease in women because the research had all been done on men. The situation is somewhat better today, with more women and minorities being included in clinical trials, but it's far from equitable.

The gender gap in the lab is very real, so much so that the National Institutes of Health recently took steps to require scientists applying for grants to include both sexes equally in animal studies. If a grant applicant wants to study one sex only, there must be "strong justification" for excluding the other sex.[10] Obviously, sex-specific conditions like ovarian or prostate cancer are exceptions, but the NIH's apparent goal is to encourage more broad-based and beneficial research on disease and treatment.

If you're not a research scientist, you may wonder why there weren't equal numbers of male and female lab rats in the first place. Why *have* more males been studied? Is it an issue of cost? Availability? In truth it happened for many reasons, including flat-out bias. When modern medical research conducted on animals began in earnest in the twentieth century, the health concerns of women, as well as ethnic minorities, were not a priority, nor did scientists fully comprehend the biological differences between the sexes. The research standards, including the use of male-only lab subjects, reflected a cultural bias that was the norm; as a result, our knowledge of certain conditions such as postpartum depression or of higher rates of some cancers in African Americans has lagged behind for generations.

Beyond bias, there was another reason why females were not included in animal studies. On a practical level, most researchers did not want extraneous variables in their experiments, and females with hormonal cycles potentially introduced inconvenient "noise" that made it harder to identify clear patterns. A study published in 1923 showed that caged female rats ran in their exercise wheels more frequently when in estrus, the phase in their cycles when they could become pregnant.[11] That research, nearly one hundred years old, contributed to a view that has persisted to this day: Females are inherently more variable than males because of their estrous cycles. What scientist wants that hassle?

When seeking to isolate cause and effect in scientific experiments, variability is indeed a nuisance. Just think back to Mrs. Danielson's second-grade science class, when you and all your classmates were each given a dried bean that you planted in a dirt-filled paper cup, and you charted the growth rate of your little sprout depending on whether you placed it in the window or in the coat closet. Mrs. Danielson gave everyone a bean and dirt in a cup. No one got sunflower seeds and Miracle-Gro. The controls were clear. In other words, there were no extra variables.

To scientists, a restless rat in heat, running in her wheel, was evidence that female lab subjects would introduce variability that could scuttle a carefully controlled experiment. Females in estrus were just "messy" — better to conduct the experiment with all-male animals and their more predictable behavior, the thinking went, to ensure straightforward and successful research. So it went for decades, and the lab was male dominated in more ways than one. In 2009 an analysis showed that male lab animals outnumbered female subjects 3.7 to 1 in physiology; 5 to 1 in pharmacology; and 5.5 to 1 in neuroscience. These aren't good statistics if you're trying to figure out why a pain medication works on a man but not on a woman. They're especially bad if you're the woman in pain.

Some scientists have pushed back against the NIH's revised inclusionary guidelines, pointing to research that shows little to no variation in how male and female animals (or cells) respond in certain lab experiments, or, conversely, arguing that the bias is intentional because it adds depth to the research, highlighting the differences between the sexes. While those may be valid points that hold true under some circumstances, the facts remain: There are many physiological and psychological conditions that deserve more study in females, such as depression and sexual dysfunction, which may actually occur more in women than in men. And that research will not get off the ground and help women if studies — from animal studies in labs to clinical trials — avoid female subjects.

Dr. Arthur P. Arnold, a professor of integrative biology and physiology at UCLA, specializes in research exploring the biological differences between the sexes. He and his doctoral mentor, Fernando Nottebohm, were the first to discover large sex differences in specific brain circuits, through their research on songbirds. (Male birds have more elaborate songs than do females, in general, which have evolved to allow them to compete with other males and attract mates. Arnold and Nottebohm discovered that males have a larger set of cells — five to six times larger — that control singing.) Arnold points to research that shows how sex differences can promote or prohibit disease in a variety of organ systems, and he believes it's essential that more females be studied, consistent with the new NIH guidelines. But Arnold has run right up against the same objections I mentioned earlier: the belief that as soon as we shine a bright light on biological differences between males and females, as his work does, we undermine a woman's ability to achieve equality with men.

Arnold has dubbed this viewpoint "far feminist" and believes that it hurts — rather than helps — women. The findings on sex distinctions and susceptibility to disease are proof that if we deny the biological differences between the sexes, we slow down advances in healthcare for women. Similarly, my findings suggest that if we deny these differences, we also lag behind when it comes to understanding women's sexuality — and possibly their intimate relationships more generally — as well as their health.

For instance, how is it that men in search of sexual satisfaction wound up with a little blue pill with a catchy name and — many years later — women got something called flibanserin, a drug prescribed to address HSDD, short for "hypoactive sexual desire disorder" (i.e., low libido)? Addyi, the brand name for flibanserin, doesn't work the same way that Viagra does, and it has less to do with physiology than with psychology.

Men take one dose of Viagra (or Levitra or Cialis) because they want to have sex within a half hour or so, and the drug increases all-important blood flow to the penis. Women take Addyi because they don't want to have sex, and flibanserin attempts to reverse this by lowering serotonin in the brain and increasing dopamine levels, to put it simply. A woman must take Addyi every night at bedtime, even if her partner is already in REM sleep, out of town, or just not in the mood. It is effective only in premenopausal women, and any alcohol use is strictly contraindicated, as it can dangerously lower blood pressure. (The no-alcohol issue could explain the fact that sales are reportedly, uh, flaccid.)

Think about it. That's true inequality.

Why isn't there a "female Viagra," and for that matter, why did it take us decades to discover the harmful side effects of birth control pills and adjust dosage to avoid them? Why don't we know more about women? It is likely that female arousal is very complex and more difficult to stoke than delivering blood flow. But, certainly, we'd know more if we studied females more. Even now, there is still a tendency among biologists to study penises to the exclusion of female genitalia. Within the last decade, half of all studies on the morphology of genitalia across species focused solely on males.[12] Less than 10 percent examined females.[13] And it's not because female genitalia are uninteresting. Some waterfowl have elaborate labyrinth-like genitals with several dead-end vaginal sacs that can function to shunt away the sperm of unwanted males.[14] The researchers documenting the penis bias concluded that it is unjustifiable — and it could reflect assumptions about the dominant role of men in sex. (It could also explain why we can't all agree if the G-spot is real or if it's a unicorn that only comes to life on the pages of *Cosmo.*)

The fact is if women don't catch up in the lab, they won't catch up in the real world.

Unlocking the Answers

For generations, we've been basing our knowledge of female relationships and health on what we have observed about males. In the sexual realm, males pursue, they compete with other males, they have larger sexual appetites — they dominate. Peacocks put on a brilliant show and the homely-looking peahen emerges from the brush. Silverbacks kill other males and mate with numerous females. Male lab rats aggressively mount females who respond with a welcoming reflex that allows for impregnation. But this is an out-of-date, limited, *Wild Kingdom* view of sexuality that repeatedly casts females in a passive role designed to make them receptive to males, a view that is at odds with the science that has emerged in the last decade, and with reality.

The way to understand female sexual behavior — from desire

to sexual response to reproduction — is to quit looking so much at male sexual behavior. Instead, we must continue to explore why women do what they do by studying their actions, not simply their reactions to males. Specifically, we need to look more closely at the role of the hormonal cycle and how the female brain has evolved to take full advantage of its distinctly timed phases. It's difficult to gauge how the bias in research might have hampered progress in the study of women's health and well-being, including female sexuality and the role of the fertility cycle in particular, but it's time to put the old attitudes aside and press ahead.

It's time to embrace the intelligence that comes with being hormonal.

2

Heat Seekers

THIS IS THE STORY of scientists in search of a previously unidentified phase of women's sexual and social behavior — a phase guided by hormones. It turns out that in order for us to discover it, we had to revise our thinking about female sexuality more generally.

The words "females in heat" stir images of lady cats prowling the neighborhood alleys and calling out loudly for randy toms to mate with, or wanton women who lose control or fall prey to desirous men.

The real idea of "heat" — the common name for the phase in the fertility cycle when pregnancy is most likely to occur — is far more nuanced than that, and as a biological phenomenon in animals and humans, it deserves to be properly explored and not trivialized as nature's booty call. It's not as simple as *Give it to me baby (and give me a baby)*. If we want to know for sure how hormones influence female sexual behavior, let's start with taking a closer look at so-called heat, or *estrus*, which occurs just before the ovaries release eggs for the possibility of fertilization by any waiting sperm. It has taken us most of human history to understand the role of estrus beyond reproduction, and we're still discovering its secrets.

In science, our evolving views surrounding estrus date back to ancient times. It starts with depictions of nubile mortals (or goddesses) driven into a state of frenzy by their feminine desire to catch and keep a mate. Though not necessarily "in heat," such women, mythical or real, were written off as hormonal in some part of their narrative: the temptress Eve, vengeful Hera, passionate Cleopatra, and a further parade of real-life femmes fatales, scheming queens, witchy women, and black widows galore. Here was an almost fearful view of womanhood unchained, which paradoxically existed at a time when women were often repressed and powerless.

Then came the second wave of thought, as modern science gained traction, and the frenzied-female archetype was overshadowed (though never entirely replaced) by a less threatening model — that of the passive female whose monthly hormonal flux made her more receptive to male advances. When her man knocked, she'd jump to answer and fling the door wide open. This was a nice and tidy way to view female hormones, which perhaps were nothing more than a way to guarantee pregnancy and the continuation of the line.

Finally, there's the contemporary version of the estrus story, in which females play a far more active role in directing their own sexuality and reproduction. They aren't the libidinous and hysterical women of yore, nor are they simply the baby-making vessels of well-timed male attention. Looking for those patterns in women's behavior turned up no new insight about how women actually did behave (and they often failed in animal studies too). A more contemporary view allowed us to discover the true nature of the relationship between hormones and female sexuality, and to discover an estrus-like state in women.

The Heat in Austin

I was a grad student at the University of Texas, studying, working, and doing what grad students typically do on hot, humid nights in Austin: I was at a party with other sweaty grad students, and drinking a cold bottle of Shiner Bock. So many warm bodies in such close confines and everyone smelled a little strong, especially *him*....

My journey as a scientist tracing the path of estrus began in graduate school, not only with the research being conducted in the lab, but with the observation of shifts in my own behavior during certain times of the month, and noticing the same patterns in my female friends. My views on men, my views on myself, my general interest in being social, changed from day to day, as did those of other women. With my background as an evolutionary thinker, I didn't find it plausible that these changes were due to some sort of feminine stereotype — the fickle female, the contrary woman, the girl who runs hot and cold — and as I continued to explore the science, I was drawn to the parallels I saw between human experience and the experiences of our animal cousins.

I'd been taught the accepted wisdom at the time: that humans did not change their sexual behaviors during the fertile phase of the hormonal cycle, and that ovulation was a completely concealed event. Nearly all other mammals, however, behaved quite differently during estrus. (Ladies didn't let it all hang out; female baboons, with their swollen genitalia, literally did.) But the more I thought about the evolutionary forces shaping humans, the more I questioned that view.

Evolution is about smart reproductive decisions. Surely, human evolution would have favored a special set of mental decisions surrounding mating that took fertility into account. For animals and humans alike, the benefits and potential costs of sexual behavior are high; successful mating results in the perpetuation of the genes that gave rise to those successful mating decisions, and failure is an evolutionary dead end. Why in the world, I asked myself, would our brains evolve to be indifferent to variation in fertility across the cycle?

Which takes me back to that night in Austin, where I was considering that very question. I have a very powerful memory of that evening. I was sitting next to a friend, a supersmart guy who loved ideas, and a fellow Darwinian like me. I think he'd ridden his bike to the party, because his body scent was strong — spicy, pine needle-y, musky — a smell that ordinarily I'd have found to be too much. But that night I thought he smelled interesting, even — could it be? — *sexy*. I glanced over at him. I had never noticed that he was attractive. Ordinarily, I'd have found his face a little too angular, too large, too masculine-looking. But something made me see him differently. I had noticed this kind of thing before, when a man I'd met initially seemed uninteresting. Then I'd see him again at a later date and wonder why I hadn't paid closer attention.

A few months later, a scientific paper would be published — the now famous "stinky T-shirt study" — which, in part, traced what women found attractive in men during certain phases of their cycles.[1] Among the findings: Fertile women are more attracted to men with symmetrical features, and scent provided a cue, leading women to particular men. Male symmetry is an important factor because it potentially indicated the presence of strong genetic material: high-fitness genes from a male that a female could pass on to her offspring, thereby ensuring its survival and its own reproductive success (at least in ancestral conditions, when humans faced the peril of disease or injury, without the benefit of a local hospital).[2]

That research made me realize this: It wasn't that I didn't pay attention to certain men; it was that what I found attractive might change systematically, in a hormonally intelligent way. Back at that party, I wasn't just sipping beers with a suddenly handsome man who gave off a surprisingly sexy scent; I was using hormonal intelligence to sniff out a potential mate — just like some animals do. That paper, which would go on to become a key part of the heat-seeking story, was the first to provide compelling evidence that human females might possess something like estrus, which our nonhuman cousins had been making the most of for generations.

Estrus: Let's Start with Animal Magnetism

Perhaps it's been difficult to see similarities between animal and human estrus because animals, unlike us, are so . . . *obvious* about their fertility (there's nothing subtle about the aroused, inflamed external genitalia of some animals in heat). Still, despite the radical differences in human and animal behavior, if we're searching for a connection, it's important to understand how this biological phenomenon unfolds across the species. (And at the very least, you'll learn some grammar: "Estrus" refers to the noun form, as in "a female in estrus," whereas "estrous" is the adjective, as in "her estrous cycle.")

Estrus is a particular phase of female sexuality, one distinct part of the larger estrous cycle (also referred to as the hormonal or fertility cycle). It takes place immediately before the ovaries release eggs for the possibility of fertilization by any waiting sperm. In rats, mice, dogs, and many other species, estrus is the *only* time females will mate, a major difference between animals and humans (and many primates, as it turns out).[3] And in perhaps all species with estrus — including us — it is the time when males find females especially attractive.[4]

Females who mate only during estrus have very little interest in their male counterparts at other times in the estrous cycle. Hamster females take this strict time frame to an extreme level of rigidity — you'll never think of that sweet-looking little fur ball, curled up in the wood shavings of her cozy enclosure, in quite the same way. The female hamster is among the most aggressive females in the animal kingdom. A female will brutally attack any male she encounters, locking her body to his in a rolling fight and rapidly biting him, sometimes using her hind legs to suddenly launch herself away, carrying a bite of his flesh in her mouth.[5]

But not when she is in estrus. When a female begins to approach ovulation and enters estrus, she exits her burrow, leaving a fragrant scent trail for males to follow. When a male arrives, the female welcomes him into the burrow and engages in sex. However, once she is done, *she is done.* She becomes aggressive again and will chase the male hamster right out the door.[6] (Pet-store owners know that if they open a shipment of hamsters that are not separated by sex, they'll find a lot of dead males — with bite marks.)

But it's one thing to be willing, another to be able. Many species, including the hamster, are only *physically capable* of mating during estrus. Hamster females have an open vaginal canal only when in estrus. Following estrus, they grow an imperforate membrane — the scientific term for "penis force field" — that walls off any possible male access to sex. (*Guys, it's not happening, and if you don't leave me alone I'll bite you to death.*)

Female rats have a reflex called *lordosis* that is under strict hormonal control; lordosis can occur only during estrus, and it's required for the male to achieve copulation.[7] When a male approaches to mate, he rubs against her hind legs and she swishes her tail aside and adopts the lordosis reflex by curving her back downward and her rear upward so that the male can *intromit* (another clinical term, but you hardly need to look it up). Without lordosis, the downward orientation of the vaginal canal makes it mechanically impossible for a male to mate with a female.

It is hormones that cause these changes in females — not just the willingness to mate but the ability to do so. Scientists who've observed these patterns over and over again are convinced that there's a direct link between estrous hormones and female sexual behavior.

Opening the Window

So, within much of the animal kingdom females have sex only during their most fertile window — a state called "classic estrus." At other times during the cycle, females aren't motivated to seek out mates and they'll refuse any attempts by males to copulate.

Monkeys, gibbons, apes — and, yes, humans — may have sex at practically any time within the cycle.[8] But even among these primates, it turns out that sex takes place most frequently during classic estrus,[9] though some have observed that chimps might be more promiscuous during nonfertile phases. (And in the case of humans, even if sex doesn't actually occur while the window is open, estrus-like sexual desires do, as you'll see later.) For as long as scientists (and many laypeople) have been paying attention, they've observed that female sexual behavior is different when females are in heat. But what, precisely, do changes in social and sexual behavior mean? And how do we interpret those behaviors if we want to truly understand "hormonal" females?

Here's the simple answer. At first glance, it appears that estrous behaviors — from hamsters leaving scent cues to women finding certain men more attractive than others — represent an increase in a female's desire to have sex. From an evolutionary standpoint, this would seem to make sense. Estrus is prime time for becoming pregnant, so the female brain is getting this news flash, courtesy of the sex hormones: *Hey, girl! The time is right, have sex! Don't become an evolutionary dead end. Get sperm to fertilize your eggs and have babies!*

There are two simple versions of this scenario. In one, females go on the hunt for males and solicit them for sex. In the other, females are more passive. They emit some attractive chemical cues and let the eager males come along. Then they're simply receptive to sex with those males. Either way, females get sperm and propel their genes into the next generation. Mission accomplished.

But it turns out that's *too* simple — too simple for animals, and definitely too simple for humans. What we're discovering is that there's something far more subtle and interesting at work in the brain that causes females to play a much more active role during estrus than previously thought. While we know this period marks a time when they're motivated to have sex, we're discovering that females are exceedingly selective. They're carefully seeking out certain kinds of males with particular traits.

Female sexuality is strategic. But we didn't always view it as such.

A Brief History of the Frenzied Female

The ancient Greek origins of the word "estrus" reveal early and persistent notions about female sexuality, whether the females in question happened to be goddesses, mortals, or animals. In the Aeschylus tragedy *Prometheus Bound*,[10] Zeus falls in love (yet again) with a nubile young thing who is not his wife — his very jealous wife, Hera. To disguise the luscious, irresistible Io from his vengeful spouse, Zeus transforms his lover into . . . a cow. (One is led to think that there must be some lonely farmers out there who have tried to convince people that their livestock are actually Greek goddesses.) When the furious Hera discovers her husband's latest infidelity, she dispatches a gadfly — in Greek, *oistros*, a fly that pesters and bites livestock. The gadfly literally bugs poor Io into a frenzy, driving her farther and farther from home, and from Zeus. *Oistros* produced a restless frenzy in Io, just as estrus was thought to drive female mammals into a wild frenzy of sexual desire.

Aeschylus used *oistros* to describe a state of madness. Later, in *The Republic*,[11] Plato would use the term to refer to "confusion," and in *The Odyssey*,[12] Homer would use it to refer to "panic." So, the modern term "estrus" has its roots in a word that connoted madness, frenzy, confusion, panic. The Greeks laid the foundation for many great things, but they also are responsible for seeding the idea that estrus caused women to act without logic, to become indiscriminate, lustful, and sluttish.

For as long as humans have lived with domesticated dogs and cats or bred livestock, they've taken note of animals in heat, particularly estrous females who seem eager to mate. The idea of female sexual heat among animals even shows up in the Old Testament.

Here's God himself, talking about camels: "When she is in heat, who can control her? No male that wants her has to trouble himself; she is always available in mating season."[13] In the time of swords and sandals, the female was considered uncontrollable and insatiable, ready to have a go with any male because she was "always available in mating season," and that idea would prove to have staying power, right through to the modern era.

By the late 1700s, English text describing animals as being "in heat" was common, and the term shows up in written records of farming procedures. We know that estrous cows bellow more and become restless (like their ancient ancestor Io); similarly, estrous pigs, sheep, and goats vocalize more and trot or walk along the perimeters of their pens, as if looking for a way out to where the boys are.[14] Estrous dogs and cats wiggle under fences or leap over them, sometimes traveling for miles in search of a mate. Estrous rhesus monkey females will press a bar more rapidly to open a door leading to a male's cage.[15] Estrous rats will cross an electrified grid to gain access to a male.[16]

All of this evidence, combined with the popular depictions of females in a state of frenzy, contributes to a picture of females getting their hormonal switches flipped during estrus — "turned on," so to speak. And once that switch is flipped, the females put the lust into wanderlust as they seek a partner, any partner, for mating — hopping fences, breaking down doors, taking painful risks, doing whatever it takes to satisfy their sexual urges.

But once again, animal behavior — and our own — is just not that simple. As early animal researchers attempted to decode estrous behaviors, this notion of promiscuous females in sexual heat would give way to another, very different theory, a drastic downshift for the female. She went from out of control to utterly submissive.

Boy Meets Girl—and She's "Receptive"

After years of observing animals in the lab, researchers began to rethink the role of the female, questioning to what extent her estrous sexuality controlled her behavior and the reproductive process.

Was a female in heat the initiator, or was she more of a recipient of male advances? Scientists were flipping their perspective to one that was quite the opposite of the one that described the frenzied female in estrus. For much of the twentieth century, they believed that females were largely passive recipients of male advances. (Though animals were the focus of their studies, in a sense their new perspective was a reflection of human society—outside the lab, men traditionally led and women generally followed.)

This males-in-charge view was fostered in part by a focus on male sexual behavior, which scientists believed was more complex than that of females. In the late 1960s, Frank Beach, a pioneer in the study of sexual behavior, conducted research that seemed to suggest exactly that (though he changed his views later).[17] He'd observed that male beagles, long after they'd been castrated, would still mate successfully with females—not because they fathered a litter of pups (impossible in their castrated state), but because they achieved the canine equivalent of something approaching a postcoital cigarette in bed. The signature of successful mating in dogs is something called a postcopulatory lock—a rump-to-rump positioning, facilitated by the male's penis, which functions to keep the couple together while seminal fluid fully transfers from the male to the female. The male is literally "locked" into the female until the process is complete. Castrated beagles can still achieve this.

However, removing the ovaries (and the resulting hormones) from a female brings her mating days to an abrupt end, as she will refuse any attempt by a male to mate.[18] Beach's research suggested that female sexual behavior was under tight hormonal control: turned on in estrus (with intact ovaries and hormones firing away), turned off otherwise. Males, however, behaved in a more complicated way that was not dictated by a hormonal cycle.

Estrous females, then, were thought to merely respond to stimulation by a male.[19] The male (rodent or canine) approached, and the female, if in estrus, was "receptive" to copulation. This passive female idea fit neatly with the paradigm of the dominant male in the animal kingdom; if one subscribed to that view, it made sense that the male would be the aggressor and initiator, not the female. And, of course, many did believe that this was nature's way, in part because females were not being studied as closely as males. Let's not forget the penis bias in animal research (Chapter 1); because females were not studied as extensively in the lab as males, scientists simply knew less about their sexual behaviors.[20]

But a new understanding of male/female sexual behavior was on the horizon. Frank Beach himself eventually began to rethink the theory of passive female behavior, particularly as he observed what happened between beagles earlier in the mating game, well before they were locked in their ultimate embrace.[21]

Boy Meets Girl, Take Two—Active, Not Passive

If males led when it came to the mating dance, then how did that explain an important observation that Beach had made in his own research? Females in heat, he observed, sought to elicit chase from a male. If a male beagle was tethered to a stake and could not chase a solicitous female, she'd lose interest and seek out another male. *What? You're not going to run after me? Well, then, forget you!* Beach thought this might indicate a more general principle in animal mating. She wasn't just being a coy, teasing female — this was a true test for the male. *I won't mate with you unless you can* prove *that you can catch me (or attract me or woo me, or otherwise prove that you are a worthy father for my offspring).* Here was proof that female dogs — even when in heat — were not seeking just any male. He had to be strong enough, fit enough, to catch her.

Interestingly, and perhaps not surprisingly, the theme of "female strategic choice" was promoted by the research of an emerging cadre of young women entering the field of biology in the 1970s and '80s. Psychologist Martha McClintock noted in her 1974 dissertation that little was known about rats in the wild, despite the fact that much of what we knew about reproduction and physiology was based on rats — in cages in the lab.[22] Consider the standard lab-rat sex scenario, where the standard female-passive pattern emerged. The male approaches and mounts the female, grabbing and palpating her hind legs. She responds with the lordosis reflex, arching her back and allowing the male to copulate with her. After several episodes of copulation the male ejaculates. He rests before returning to repeat the pattern.[23] After multiple mounts and ejaculations, the female — no surprise here — gets pregnant. So, in the lab it was *male approaches, female responds.* Wham, bam, thank you, ma'am.

But in the wild, rats engage with one another very differently. What McClintock showed in her work, along with her contemporary, biologist Mary Erskine (a pioneering neuroscientist considered an expert in rat behaviors), was that females in their native environments do not simply receive the advances of males, as they do in the lab. Female rats typically live in snaking burrows with multiple males and females. Consider that for a moment. In the lab, rats are often separated by sex; they may even be in a rodent version of solitary confinement. When a male and female are housed together, a female doesn't get to pick her mate — she just gets assigned one. If the social structure is upended in a basic lab setting, it follows that behavior might be altered as well.

In the wild, because females live in close proximity to males in these labyrinthine burrows, they have the ability to approach males or run away, as well as the opportunity to pace their sexual behavior by choosing the sequence of males they'll mate with.[24] Instead of living in a girls-only dorm under bright lights that illuminate every action, they have the run of a nightclub with lots of dark corners. And they do what they please.

McClintock's research showed that copulations in wild female rats are preceded by distinct female behaviors. First the female approaches the male of her choosing; then she runs by him to further attract his attention, wiggling her ears, hopping, and darting.[25] This is drastically different from lab-rat sex, where the male approaches from behind and then simply hops aboard. In their natural environments, females mate multiple times with different males, and it turns out the males that ejaculate first and last in the sequence of matings father the most pups. It seems female rats in the wild are playing a very active role in mate selection, determining which males (the first and the last) will pass their genes on to offspring. These males might be the most dominant, which could be correlated with their offspring's health or future reproductive success. Or, possibly, they possess genes that will combine well with the genes of the female, which might result in fitter offspring (see the discussion of MHC genes in Chapter 6). A caged female in a lab setting has no choice in mate selection (or if she'll mate with more than one male), and perhaps little choice but to submit to male advances; fighting off another animal might simply be a futile waste of energy or could prove to be dangerous.[26]

This same pattern of female strategic choice is echoed in more recent research showing that estrous females prefer and seek out dominant males. In one field study, Italian biologist Simona Cafazzo led a team that followed a pack of wild dogs living in the streets of Rome. They found that females in heat searched for high-ranking males and mated more frequently with them, contributing to a greater number of puppies sired by those particular males.[27]

So, estrous mammals such as rodents and canines are more discriminating than we once thought. But what about some of our closest primate cousins, chimpanzees and orangutans? Though the evidence is somewhat mixed, it appears that they prefer high-ranking males when in estrus, once again showing that females attempt to exert some control over whom they choose to father their offspring. Wild chimpanzee females that display maximum sexual swellings (a hard-to-miss physical display that indicates impending ovulation) engage in repeated matings with high-ranking males, more than at any other time in their cycles.[28] We can't rule out that those males simply scare off the competition and exert their will when females are most fertile, but I am willing to bet on the active choice of female chimpanzees, too.[29]

Female orangutans in estrus also engage in strategic sexual behavior, likewise favoring dominant males. Not only are dominant male orangutans physically larger than nondominant males; they also have a distinctive feature that marks them as dominant — large, fleshy cheek pads called flanges, which may be associated with higher testosterone levels.[30] Females will mate with nondominant (less cheeky!) males too, but when they are in estrus, nearly all their matings are with the dominant, flanged males.[31] Again, I am willing to bet that they're choosing the fathers of their offspring, and they like those big, floppy faces.

Among chacma baboons, sexually swollen females form "consortships." Consortship may sound like a quaint-sounding ritual (Queen Victoria and her beloved prince consort, Albert) or a legal term, but it is yet another form of strategic sexual behavior on the part of a female. A consortship among chacma baboons is when fertile females sit close to, groom, and mate nearly exclusively with high-ranking males.[32]

Remember our estrous lab rat from Chapter 1, who ran endlessly in her wheel back in the 1920s? Let's consider why our girl might have been taking her run to nowhere.

As the research eventually would show, she, like other mammals, including primates, wasn't a frenzied, panicking female in heat, restless and mad with sexual energy. Nor was she the passive recipient of sexual attention, the wallflower waiting for any male to approach her. If she'd been in her natural habitat, she would have been moving around freely, seeking out a fit father for her offspring and making the approach. It was a mating dance, but it was almost always a Sadie Hawkins Day dance.

During estrus, it's clear that the female actively solicits sex with a male of her choosing. She tests him to see if he is fit enough to follow. She shows a preference for a particular type of male.

She picks *him*.

That's how it works in the real world. Back in the lab, it was (and still is) different. At least our girl had the wheel so she could stay busy — the guys in the other cage, the randomly assigned guys, were simply too boring.

There's a fascinating historical irony in the story of the search for the strategic female exercising *her* desires. Charles Darwin proposed it many, many years before — in 1871 in *The Descent of Man, and Selection in Relation to Sex.*[33] For Darwin, female choice was required to explain extravagant male displays, such as the peacock's tail. Darwin puzzled over why the female would be so persuaded by male aesthetics, but he suspected it was not just because those boys were so colorful. Their pretty features conveyed something more, perhaps something useful for a female to transmit to her offspring. And perhaps the same could be true for human females.

What about Us? The Search for Human Estrus

Early in the twentieth century, physiological scientists assumed that human estrus existed. After all, "heat" was well documented among animals, in and out of the lab. But up to now their predictions about mating patterns had proven to be too simple and mostly failed.

Still, it seemed reasonable to mid-twentieth-century scientists that an estrus-like state would exist in humans. By then, scientists had already figured out that they could gain insight into the physiology of women's menstrual cycles by observing the cycles of other animals.[34] Rats, as it turns out, have a hormonal pattern that is strikingly similar to that of humans.[35] Unlike humans, however, rats afforded the opportunity for experimenting.

Scientists could remove hormones or replace them and observe the effects. For example, estradiol, one hormone in the family of estrogens, peaks just prior to ovulation. Experiments on rats showed that estradiol played a key role in female sexual response — remove it, and you remove the all-important lordosis reflex necessary for copulation and reproduction; replace it, and the reflex and subsequent sexual activity return. Given the physiological similarities between humans and rats, it was sensible to think that behavioral patterns would also follow suit.

Based on that logic, a seemingly straightforward prediction would be this: Women will have sex more often when most fertile, when estradiol levels are high and the chances of conception are at their peak. The fertile phase of the cycle includes the day of ovulation and several days before it, when sperm remain viable in the reproductive tract waiting for ovulation to occur.[36] For women with a typical cycle length of about twenty-eight days, those fertile days begin about eight days after the onset of their menstrual period. (Very roughly, the fertile phase occurs about midway through a woman's cycle. For more details on the hormonal cycle, see Chapter 3, "Around the Moon in Twenty-Eight Days.")

In the late 1960s, researchers began to test these predictions, with the first systematic study of the timing of women's sexual intercourse and orgasm across the cycle.[37] Each day for ninety days, women filled out slips of paper and delivered them to a collection box at the researchers' lab at the University of North Carolina. On these slips the women indicated whether they'd had sex and whether they'd had an orgasm. There were several findings. First — and least surprising — women had more sex than orgasms. Second, and more noteworthy, on average there were two peaks in the frequency of both intercourse and orgasm: one around the middle of the cycle, which would be roughly consistent with estrus; and, more mysteriously, another just prior to menstruation.

Third, there was a great deal of variation among women in the pattern of sex and orgasm throughout the cycle, which made it difficult to draw any solid conclusions from this first-of-its-kind study. Some women showed peaks midcycle. (Was this classic estrus, when sex is reserved for the fertile window?) Some women showed consistent — or flat, as in low — rates across the cycle. (Did this indicate something about the quality of their relationships — or partners?) In a second study, about ten years later, the authors would find considerably more random rates of intercourse and orgasm across the cycle, raising more questions than answers.

Keep in mind that this research was taking place in the late 1960s and '70s, and asking ordinary women to be blunt about their frequency of sex and orgasm, even in the name of science, was still a new idea. The findings of Masters and Johnson were just beginning to see the light of day, and Helen Gurley Brown's "*Cosmo* quiz" sex surveys had yet to become must-reading. (Brown took over in 1965, and her October issue featured this cover line: "Choose the Perfect Father and the Sex of Your Baby." Did the future author of *Sex and the Single Girl* know something about female strategic choice that researchers did not?)

Later studies would again show that women had sex more frequently during their fertile phase;[38] in some cases, the increase was found only for sex the woman initiated.[39] In one study, there was a decrease in female-initiated sexual activity during the fertile phase, but an increase in women's "autosexual behavior" (in other words, masturbation).[40] The participants seemed to be as varied as the findings, with many studies including small numbers of women (as few as thirteen),[41] as well as a mix of married and unmarried university students, who might not be representative of female humanity at large. So, any general patterns of change in women's sexual activity across the cycle were hard to see.

However, there was one remarkable study that did seem definitive, given its massive scope. It included more than twenty thousand women across thirteen countries. The finding: not even a hint of an increase in women's rates of intercourse on fertile days of the cycle.[42] The conclusion: Women's sexual activity is certainly not under any sort of strict hormonal control. This is where the animal/human parallel diverges; women, unlike females of other species, were fully emancipated from the control of their hormones.

But science rarely allows a conclusion to sit for long. Here, then, is the follow-up question: If a woman's sexual activity is not directly regulated by hormones, what about her *desire* for sex?

Maybe It's the Thought That Counts

Sexual behavior is constrained by many factors, including having a willing partner and having time to do the deed. Case in point: One of the most robust patterns in sexual behavior is the "weekend effect" — about 40 percent of the sex couples have is on the weekend.[43] Unlike rats lolling about in their laboratory cages, people are busy — going to school or work, taking care of themselves, taking care of their kids, and just going about the business of everyday life.

It's not always convenient to have sex, even if you are feeling a hormonal nudge and you've penciled in "date night" on the calendar. So, perhaps the best place to look for estrus-like changes is not in human sexual activity but in thoughts and feelings about sex, which are arguably far less constrained by the demands of daily life. Here, too, though, the findings are mixed. Some studies find that a woman's desire to have sex is greater on fertile days of the cycle, but others find peaks just prior to menstrual onset, or, again, no clear patterns.[44]

In research in my own lab at the University of California, Los Angeles, we have found no evidence of increases in women's feelings of general sexual desire on fertile days of the cycle, despite using sensitive methods that follow women over time and verify cycle phases using hormone tests.[45] But a hundred miles up the coast from my campus, at our sister school UC Santa Barbara, my colleague psychologist Jim Roney, also using rigorous methods, has found different results in his lab — a general *increase* in women's sexual desire on fertile days of the cycle. (One wonders whether there might be something about the two sets of male students women are thinking about — UCLA versus UCSB — when they are reporting on their desires. . . . Are studious guys in button-down shirts less appealing on those estrous days than guys with surfboards?)[46]

The simple prediction that women would be more motivated sexually — in thought and in deed — during the fertile phases of their cycles has not been resoundingly supported by the research, despite what we've observed in animal behavior. But does this mean that an estrus-like state does not exist for humans? I think it means that we should be asking a different question: Is there a reason that women have such a variety of behavioral patterns across the cycle?

The answer, in my opinion, is yes. And the reason may be that there is an underlying strategy to a woman's hormonal behavior: Not just any male will do.

Heat Seekers, Still Seeking

On the surface, the results of all these studies could lead us to conclude that there is undoubtedly no human estrus if the objectives were just to find close parallels between hormonal human and animal behavior. But what if we were looking for the wrong pattern? If so, researchers who gave up did so prematurely.

In the early 1970s, researchers showed that women did the equivalent of rats running in the wheel on fertile days.[47] Fifty years earlier, as discussed in Chapter 1, scientists established that female rats used their wheels the most during estrus. Women in the study wore pedometers each day for three or more complete ovulatory cycles and went about their daily lives. There were actually three peaks in their "locomotor" activity. On average, women walked more midcycle, paralleling the rat-study results, but also right at the beginning and end of their cycles, showing that the human pattern wasn't exactly the same as the rat's.

Still, this was provocative evidence of real behavioral changes across the cycle. And that's what is worth focusing on here; it's not necessarily *how* behavior changed and if it paralleled that of the rats; it's the fact that it *did* change at all, because it did so for a reason. So then, *why did it change?*

In another pioneering study, also from the 1970s, women used standard cotton tampons inserted overnight to collect vaginal odors, over the course of fifteen ovulation cycles.[48] The samples were frozen until ready for "sniffing sessions."

In those sessions, both men and women reported to the laboratory to smell the samples in glass jars. Samples collected near ovulation were rated as more attractive than samples collected during any other point in the cycle.[49] Many animals emit odors that attract mates; these scent cues marking the fertile period of the cycle are among the most ubiquitous across the animal kingdom. (And, of course, let's not forget the sway of a stinky T-shirt — but more on that a bit later.)

Both of these studies yielded enough promising information for scientists who wanted to continue the search for human estrus. But it was not until the late 1990s that the most compelling evidence of human estrus appeared, and research in the emerging field of evolutionary psychology in the intervening years would help to lay the groundwork.

Breakthrough

The study of human social behavior — why we do what we do with others — has long been dominated by a set of guiding assumptions: Humans are vastly different from other animals; humans possess no animal-like instincts (or very few of them); and human behavior is all learned and therefore culture-specific.[50] In the 1980s, evolutionary psychologists began to challenge these ideas. Evolutionary psychologists believe that Darwin's theory of evolution should be applied not just to human physiology but to psychology as well — certainly our bodies adapted, but so did our brains, and along with that our thinking and behavior.

Our brains evolved to be problem-solving machines, addressing the challenges our ancestors faced, such as finding nutrient-rich foods, finding shelter, and of course, finding mates and making babies. One of the fascinating observations that spurred evolutionary psychology is that our brains seem to be geared toward the Stone Age past and not necessarily the present. Case in point: Humans' fears and phobias center on snakes, spiders, and other creepy-crawlies, but not necessarily on our true sources of peril in the modern world. No one ever has recurrent nightmares about damaged electrical outlets, elevated blood-sugar levels, or cars traveling ten miles per hour above the speed limit, but these things are much more likely to harm us. So, the legacy of the ancestral past seems to be locked into our modern human minds, and, the argument goes, we need to have an understanding of our evolutionary heritage to fully understand our minds and behavior.

Some of the early work in evolutionary psychology was inspired by a theory developed by the evolutionary biologist Robert Trivers, widely regarded as a genius and one of the fathers of our modern understanding of social evolution. Trivers's theory, known as *parental investment theory,* states that there will be biologically based differences between the sexes when it comes to reproduction.[51]

The ideas are based on simple economic-biological principles. The sex that is required to invest more time and effort in producing progeny, and that is more physiologically limited in how many offspring it can produce, will be the most selective in choosing a mate. The sex that invests less, and that can potentially produce many more progeny, will compete for access to the "bio-economically" valued high-investing sex. Or, to put it another way: *Attention, females — tag, you're it.*

In mammals, it is easy to see that it's typically the female who invests more — the energetic requirements of producing eggs (larger than tiny sperm), maintaining a pregnancy, and producing milk. Males, on the other hand, might contribute only their gametes — sperm (though they often do contribute much more). Indeed, across mammalian species, females tend to be the choosier sex.[52] Males, in contrast, tend to be the sex that is more competitive for access to the other sex — and are less choosy about their partners.[53]

Applying this theory to humans[54] — choosy women, less discriminating men — is very controversial, in part because it violates the standard assumptions that guide the study of human behavior. (We're different from other animals, remember?) But there is actually a huge amount of data supporting these ideas. One somewhat infamous set of studies took place at Florida State University several decades ago. In the studies, male and female "confederates" (that is, accomplices of the experimenters) approached fellow undergraduate students of the other sex and said, "I have been noticing you around campus. I find you very attractive." Then they made one of three requests (randomly assigned): "Would you..." (1) "go out with me tonight?" (2) "come back to my apartment with me tonight?" or (3) "go to bed with me tonight?"

The results were stunning. Three-quarters of the men agreed to go to bed, whereas not a single woman said yes. Women were somewhat more likely to agree to go to the confederate's apartment (6 percent in one study and 0 percent in the other), whereas most men again agreed to the request. There was a fifty-fifty gender split for consenting to the date. So, women were not uninterested in the confederates; they were just not at all inclined to go for sex with a man they'd never met before. Not so for the guys.[55] This study has been repeated with similar results many times, including recently in Denmark and France, cultures that are considered some of the most sexually liberated on earth.

One of the largest studies of sex differences included more than sixteen hundred participants from around the globe. In this study, echoing others, men were more interested in "sexual variety." When asked how many partners they would like to have over the next thirty years, men across all cultures reported that they wanted more partners than women did, typically a twofold difference.[56] And there are many, many studies along these lines. Women, more than men, appear to require more information about a possible mate before consenting to sex.[57] Men, more than women, are more eager for short-term sexual opportunities.[58] Men, more than women, compete for mating opportunities,[59] and so on.

Let's look at how those behavioral patterns could square with the existence of human estrus. In parental investment theory, the prediction is that women (the high investors) will be very selective when choosing sex partners. So, the notion that a woman in estrus would become *generally* more interested in sex because it's that time is problematic. It makes little sense for a woman to pursue or accept just any man when she is fertile and the probability of conception is at its peak. The fact is, she's about to put in a whole lot of work here, and she needs to choose her partner carefully. Therefore, men who can contribute to the success of a woman's hard-won offspring are the ones we predict she'll seek.

What do women look for when they're on this seek-and-find mission? One possibility (which we'll explore in depth in Chapter 4) is that they have evolved to seek men with characteristics indicating that they possess high-quality genes — genes, for example, that might confer good health or make offspring attractive so that they will themselves be able to compete for high-quality mates.[60]

One such quality is bilateral symmetry — the extent to which the two sides of one's body match. This signifies that the genetic blueprint for developing the body was executed with few flaws, which itself indicates a lack of genetic mutations and an ability to withstand the entropic forces of nature that could derail normal development (such as a contagious disease, illness, or injury), leaving the body off-kilter here or there. And so, we have arrived at the rationale for the stinky T-shirt study.

Symmetrical, Smelly, and Sexy

Psychologist Steve Gangestad and biologist Randy Thornhill at the University of New Mexico had been studying links between men's symmetry and their sexual behavior for nearly a decade. They had found, for example, that men who were more symmetrical were rated as having more attractive faces (though results here have been mixed), had more sex partners, and, fascinatingly, were more likely to report having been the affair partner of a woman (e.g., her chosen partner in an extramarital affair).[61]

Certainly, it was possible that women could detect symmetry visually, just by looking a man over. But most deviations from perfect symmetry are probably too subtle to detect in that way. Gangestad and Thornhill thought that a man's scent could also be a good indication of symmetry and the underlying qualities (good physical condition, good genes) giving rise to it.[62] Why scent? Because research shows that how a man smells to a woman is very important in evaluating a potential partner — if he smells good, he's sexually appealing; if he does not, it's a deal breaker. (Yes, this is why there is a billion-dollar men's fragrance industry.)

They furthermore reasoned that women might particularly prefer the scents of symmetrical men when the chances of conception were high — and women could pass along genes underlying those symmetrical features to their offspring.[63] Here's what they found.

In their study, forty-two men had measurements taken for things like the widths of both wrists, lengths of both earlobes, lengths of their fingers on both hands, and so on, allowing for a calculation of their body symmetry. The men went home and washed their bedsheets with lab-provided unscented detergent. They wore no artificial scents (including deodorant), nor did they eat strong-smelling and strong-tasting foods like garlic or lamb. They put on clean white T-shirts given to them by the researchers to wear overnight for two nights. They avoided cigarette smoke and alcohol and abstained from sex or sleeping with another person while wearing the shirts. After the second night, they returned their shirts to the lab in special plastic bags.

An hour after the men returned the shirts, fifty-two women reported to the lab and circulated from station to station where the shirts in bags were placed. Each woman gave every shirt a hearty sniff and rated how sexy the garments smelled. On their way out of the lab, the women reported information about their menstrual cycles that allowed Gangestad and Thornhill to calculate their positions in the cycle.

The result: Women in the high-fertility phase of their cycle rated the scents of symmetrical men as sexier and more attractive than the scents of less symmetrical men.[64]

This was a stunning finding. The prediction Gangestad and Thornhill made was so subtle. The phenomenon they documented, if it was real, was far off the radar of anyone's conscious awareness. If the finding could be relied on, it meant that women, like their mammalian counterparts, had preferences for features in mates — piqued at high fertility — that might function to transmit high-quality genes to offspring. Here was the evidence that female strategic choice, which had been demonstrated in rodents, canines, and monkeys, was a human behavior as well. It also meant that we needed to understand the role of biology — and hormones in particular — in women's sexuality.

The stinky T-shirt study was followed the next year by a similarly remarkable finding. In that study, women in the fertile days of their cycles appeared to prefer images of men who had more masculine versus feminine facial features — broader jaws, larger chins, and generally a more chiseled appearance — particularly when they evaluated those men as sex partners rather than as long-term mates.[65] Remember the female orangutans' preference for floppy-faced males? Just as orangutan females preferred males with facial flanges when fertile, perhaps human females preferred broader jaws when fertile. (It turns out that the story of women's interest in masculine faces is more complicated than that of preferences for the scent of symmetry and is challenged by recent attempts to replicate it, but the general theme — that women prefer masculinity, in the form of masculine *behavior* and more masculine *bodies* — may hold true.)[66]

The findings of both of these landmark studies led to follow-up studies and generated huge interest in the scientific community.[67] Women appeared to prefer certain male features when fertile within their cycles. Now hundreds of studies are showing that hormone cycles influence women's bodies, brains, emotions, preferences, and relationships. The findings at once confirm the deep hormonal similarities across all female mammals and the unique sexual psychology of women.

Human estrus — a heat-like state that has been observed in animals for thousands of years — is real.

3

Around the Moon in Twenty-Eight Days

WHETHER WE'RE TALKING ABOUT humans or animals, the estrous cycle is built on a foundation of key hormones that rise and fall in a carefully ordered sequence.

Women, unlike most mammals, menstruate as part of that cycle — only primates, bats, and elephant shrews have menstrual periods. But with a twenty-eight-day average fertility cycle that begins in puberty and lasts till menopause, the modern human female ovulates (and menstruates) more than any other species by far — approximately four hundred cycles over the average life span. It would have been fewer for ancestral females, who were likely to have been pregnant or breastfeeding for much of their reproductive lives. But it is still a tremendous amount of hormonal fluctuation, over and over again.

We know that these patterns of hormonal shifts impact a woman's body and brain — PMS symptoms, cramping, and menstrual bleeding aren't news. But women, unlike other species, are "undercover ovulators"; although there could be advantages in many species for keeping the exact timing of ovulation a secret, it is likely that the human female has evolved to be particularly circumspect in this regard. We'll delve into why that's the case in Chapter 6, but here's a preview: When one considers the hard and brutal reality of life in early human history, perhaps it was safer for ancestral women to keep unwanted males and female rivals at bay during the fertile window, thereby reducing risk and increasing chances for women to make their own choices about who would father their offspring. Concealed ovulation, then, is a form of hormonal intelligence.

Day to day across the cycle, a woman's outward appearance won't vary dramatically, but inside, the female body is experiencing some phenomenal physiological changes. We know these dramatic hormonal shifts can have a psychological impact as well, causing certain behavioral changes — including female strategic behaviors, as the research on human estrus has shown. But before we continue to catalogue those behaviors and the reasons behind them, it's important to fully appreciate what's going on inside the body. To understand estrous behaviors, it's useful to understand the overall estrous cycle. So, let's geek out for a moment, as it's been a while since Biology 101.

Inside Out: The Cycle Explained

"When was the date of your last menstrual period?"

That question is so frequently posed by healthcare providers that it has its own shorthand in medicine — the LMP. Many girls and women can't answer the question without consulting a calendar or counting on their fingers. But now all one needs to do is tap the period tracker app on one's smartphone and, voilà, May 3 it is!

A Note on Cycle-Speak: Menstrual? Fertility? Estrous?

You will hear people use a variety of terms to refer to the twenty-eight-day fertility cycle: For example, most doctors and scientists will refer to the human cycle of hormones, egg maturation and release, and the ultimate onset of menstrual bleeding as the "menstrual cycle." This usage focuses on the menstruation that starts each cycle anew and makes humans different from nearly all of their nonhuman cousins. For some biologists, "estrous cycle" is for the birds (or at least for the nonhuman mammals)—and for all the other nonhuman species with reproductive cycles that are influenced by hormones.[1] The problem with two sets of terminology is that it stops us from seeing common threads that link the behavior of animals and humans. Perhaps that's one reason why it has taken us so long to see that human estrus is real and that it's the key to understanding ourselves and our sexuality. To stay neutral here in this chapter, I'll just call them "ovulation cycles."

Despite what your sister told you when she was in her new age phase, the moon and your cycle have little to do with each other beyond their length – twenty-nine and a half days for a lunar cycle and slightly less for an average ovulation cycle. The moon may regulate high and low tides, but it doesn't draw menstrual blood in and out, nor will you necessarily feel lustier on the night of a full moon, though it certainly is a romance-inspiring vision. Maybe if you feel like howling and you're transforming into a werewolf, you could blame the moon for manipulating your hormones — all that crazy, instant facial hair would indicate a megablast of testosterone. But otherwise, no — it's not the moon calling the shots; it's your hormones and your brain.

There are numerous hormones that influence the cycle, but we're going to focus on the Big Five. Think of these hormones as molecular messengers with very specific instructions, traveling throughout the body with blood as their highway and seeking out specific receptor cells where they'll deliver their marching orders.

• Estrogen: *The Big One.* Estrogen fires up some major events across the cycle, and it regulates other hormones as well. Estrogen is largely what makes a woman female — causing breast and overall body-fat development (aka womanly curves), as well as cellular changes in the vagina and uterus. "Estrogen" is actually a supercategory that includes three different "estro" "gens" (as in generators of estrus). Estradiol is a major hormone that generates estrus. Other estrogens are important for understanding pregnancy and menopause, so those become important later in our story. But when people refer to estrogen, they usually mean estradiol.

• Progesterone: *The Double Dealer.* Progesterone works closely with estrogen but has its own peaks and valleys. It gets the uterus ready for pregnancy but it also wears another hat: It can make it harder for sperm to pass through the cervix during nonfertile times, keeping them out of the deep recesses of a woman's body when perhaps all they would do is harm (e.g., by carrying disease).

• Follicle-Stimulating Hormone (FSH): *"Stimulating" Is My Middle Name.* FSH is responsible for the maturation of the egg-containing follicle, among other things.

• Luteinizing Hormone (LH): *The Bungee Jumper.* LH is necessary for ovulation and loves to spike really, really high midcycle to get things going. If you are trying to get pregnant and purchase an over-the-counter ovulation test (which will probably require you to pee on the proverbial stick), you'll be looking for your LH surge. Once you find it, one or two days later, on average, you ovulate.

• Gonadotropin-Releasing Hormone (GnRH): *The Stage Manager.* Think of GnRH running instructions and interference between the brain and the ovaries. (*Five minutes to curtain! This is not a rehearsal, folks — it's twenty-eight days of production and then we do it again!*)

The ovulation cycle is generally described as having two distinct phases — the *follicular phase,* which starts on Day 1 (the first

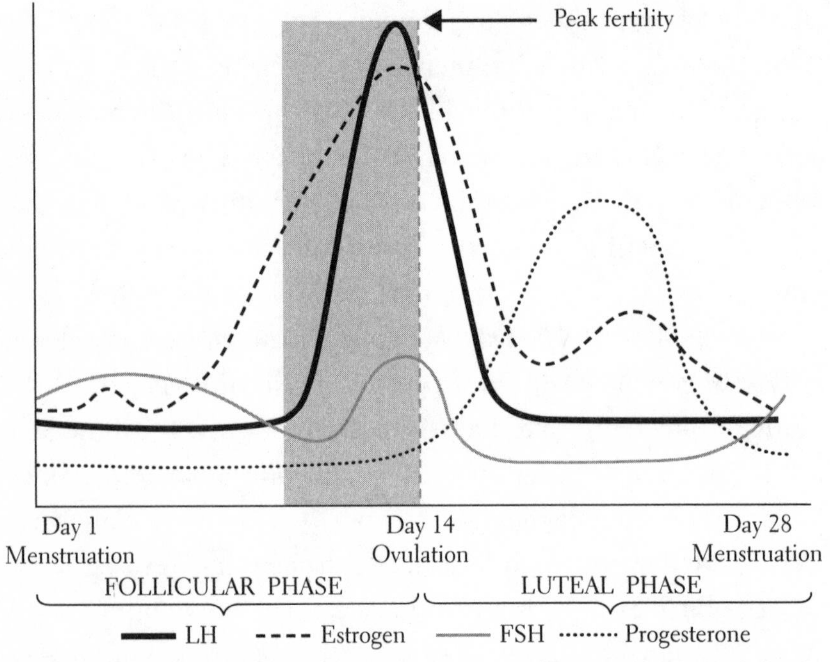

This chart approximates the phases and dates along with the rise and fall of specific hormones. Note: Although some women experience cycles that are considerably shorter or longer, twenty-eight days is the average cycle length, and five to seven days is the average length of a menstrual period. Because most women fall into the twenty-eight-day range, the cycle is usually described with that average in mind.

day of a menstrual period), and the *luteal phase,* which is the second half, from ovulation (on or about Day 14 or 15, assuming a twenty-eight-day cycle) until Day 28. Because estrogen and progesterone are flatlining from the end of the last cycle (into menstruation), *follicle-stimulating hormone* is going to step in to get the hormonal party started again, and *luteinizing hormone* will join in momentarily. Cue the glands — the dance begins.

Follicular Phase—Day 1 through Day 14

The hypothalamus, in the brain, controls the nearby pituitary gland (made up of the anterior and posterior pituitary glands). These two organs work together to get things rolling, hormonally speaking.

The almond-size hypothalamus releases *gonadotropin-releasing hormone.* GnRH is now on the move, directing the anterior pituitary gland to secrete FSH (and a bit of LH, with more to come later). Amazingly, GnRH is released in perfectly timed pulses — once per ninety minutes during this early follicular phase, and a bit more frequently as the cycle progresses, with regularity being key. It's this perfect timing that allows FSH and LH to do their jobs just right, if pregnancy is the goal. If something throws off this clockwork, like poor nutrition, illness, exposure to artificial hormones, or too much stress,[2] the system can fail.

FSH enters the bloodstream and heads south, to the ovaries, where it earns that middle name, stimulating the individual egg-containing follicles within an ovary to grow. (It could be the left ovary; it could be the right one; perhaps this is the body's way of regulating the internal workload — only Mother Nature knows which one gets chosen.) A number of existing eggs come under the influence of FSH and begin to develop, but only one lucky egg is going to *fully* mature and move on to the next round; the rest are doomed to atrophy and die off. But we're getting ahead of ourselves. . . .

As the maturing follicle develops in the ovary, it begins to secrete the *estrogen* hormone. Powerhouse estrogen goes to work, causing the cells of the uterine lining (the endometrium) to thicken and prepare to receive a mature, fertilized egg. But estrogen also calls up to the brain for some backup, as ovulation nears, signaling to the hypothalamus, which sends out more GnRH.

GnRH directs the anterior pituitary gland to crank out an extra-large dose of luteinizing hormone, which up until now has been kept in check while FSH was busy babysitting the follicles. LH is about to take its star turn, because it's largely responsible for the next event, *ovulation*, which takes place midcycle, at about Day 14.

This late follicular phase — right before the egg is released — marks the beginning of the most fertile period in the cycle, and it will last only a few days.

Luteal Phase: Day 15 (or so) through Day 28

As ovulation approaches, LH levels do a steep upward climb within a short period of time, resulting in more estrogen production, and estrogen causes even more LH to be released. These hormones, along with a steady dose of FSH, trigger the next big change — the bursting of the follicle and the release of a now mature egg (remember, only one, with rare exceptions that produce fraternal twins, will make it out of the ovary; the rest will atrophy and die). The egg sets out to meet its fate (or not — more on that in a moment), and the now-empty follicle becomes the corpus luteum (literally, yellow body).[3]

The corpus luteum gets to work pumping out *progesterone* and, yes, restarts a little bump up in estrogen. Why? Because those hormones are fluffing up the uterine lining, stimulating the endometrial cells to plump up with nutrient-rich blood and prepare for a fertilized egg.

Back to our one and only egg — it has passed out of the ovary and is making its way to the uterus through the fallopian tube. It has reached the fork in the fertility road, and one of two things will happen.

If our egg meets up with "the one" — a viable sperm — then it's bingo: fertilization, which takes place in the widest part of the fallopian tube. Because sperm can live for up to five days inside the body, sex that takes place just before ovulation can still result in a pregnancy. An egg, however, needs to be fertilized within twenty-four hours. If the egg is fertilized, it will burrow into the uterine wall, once it travels completely out of the tube, for fetal development to take place. Nine months later (actually at a little closer to ten — the doctors have not been telling us the whole truth!), collect the door prize to end all door prizes. During pregnancy, the production of estrogen and progesterone continues, while FSH and LH are inhibited, thereby preventing ovulation during pregnancy. (We'll talk more about pregnancy in Chapter 7.)

If the egg does not have a game-changing encounter, then when it reaches the uterus, it flips on the light switch and orders everyone out. This hormonal dance party is over. The thickened uterine lining has no purpose and is shed through menstruation; the first day of menstruation is considered the first day of the new hormonal cycle. The hormones have done their heavy lifting for now, and levels of estrogen, progesterone, FSH, and LH begin to recede and level off — and then the cycle starts over.

The Myth of Menstrual Synchrony

"Menstrual synchrony" is a popular idea that has its roots in a heavily circulated study about the timing of cycles among groups of college dorm mates.[4] It made its way into just about every women's magazine, dorm, and girls' slumber party. *Oh my god, I got my period too!* Men became convinced that their teenage daughters and wives were on the same cycle and used this as an excuse to take extended fishing trips. Girls and women took comfort in thinking that even if they got caught without a tampon or pad, certainly a roommate would have a spare.

This perfect hormonal harmony isn't real. Better and more recent studies do not find that women living together synchronize their cycle phases,[5] and part of the misunderstanding may be due to the variability of cycle length. But first let's look at why this phenomenon seems implausible in the first place. For starters, it doesn't simply mean women get their periods at the same time; menstrual synchrony means *all* phases of women's cycles line up, including any effects on sexual and mating behaviors.

If females living together had synchronized cycles, they'd be at peak fertility at the same time. Consider the implications for ancestral women; those without steady or satisfying partners would have been forced to compete for the same men at the exact same time. But more important, what would be the benefit of synchronizing in the first place? The underlying physiology would be complex and metabolically (and reproductively) costly. A woman would first have to detect the cycle phase in her close friends and family and then adjust her own hormone cycle, perhaps shrinking her follicular phase (giving an egg insufficient time to mature) or rushing her luteal phase (perhaps giving her uterus less time to receive a fertilized egg or rejecting an implanted egg that had made it to the womb). If all women cycled together, this would make it easier for outsiders to detect fertility—whereas it makes sense for women to keep signs of fertility on the "dark side of the moon." In short, a good evolutionary rationale for why human females would have this complex strategy with its costly underlying physiology is just not there. (Perhaps there is a rationale for some other species in which ovulation is more clearly on display. Synchrony can allow for greater female choice of the fathers of their offspring because the pushiest males cannot dominate all of the fertile females at once.[6])

The reason that it's so easy to think that menstrual synchrony exists in humans is because "normal" cycles among a group of women can easily overlap—and appear to converge. For instance, let's imagine four females sharing an apartment. Roommate A has a twenty-eight-day cycle; Roommate B has a thirty-two-day cycle; Roommate C has a twenty-five-day cycle; Roommate D never keeps track and also never remembers when the rent is due, even though it's the same time every month. Roommate A ovulates on or about Day 14, B about Day 16, C about Day 12 or 13, D can't tell you. All of them have regularly occurring periods, but some menstruate for two to three days and others closer to a week. A's third and final day of menstruation happens to be Day 1 for B, and it's also the day before C gets her period. Meanwhile, D is convinced she's getting her period any day now.

Do you see where this is going? It's inevitable that phases of the cycle will overlap at some point. Moreover, the farther away women's cycles are *initially* from one another, the more they will appear to sync up. There's only one direction to go: closer—just due to chance alone. (This phenomenon is called *regression toward the mean,* which is well-known to statisticians and can explain a variety of illusory phenomena, such as when the economy bottoms out and then is apparently improved by some new policy or leader.)

There is simply not solid evidence that women somehow synchronize their hormonal activity when they're living in close proximity, and there's no good reason why we would have evolved to do so. If you ever hear a husband, father, or brother complain about all the women in the household acting "hormonal" at the same time (and hogging the bathroom), you can safely say that the only time women truly cycle together is when they're on a group bike ride (or by random chance).

Twenty-Eight Days of Female Strategic Behavior

For most healthy women, the fluctuations in hormonal activity across an average cycle are fairly predictable from month to month, though of course they are dramatically altered if pregnancy takes place. Eventually, the cycle will be transformed with age as perimenopause and menopause approach, impacting the variety and amount of hormones that are normally released.

Beyond these internal changes, it's well documented that a woman's outward behaviors also change across the cycle. Human ovulation is concealed (*mostly;* see Chapter 6), but certain behaviors occurring before, during, and after ovulation may be on display. Perhaps the most well-known manifestation of hormonal behavior is premenstrual syndrome, though PMS wasn't widely discussed as "a thing" until the 1980s, when it was increasingly acknowledged by the medical community (and piles of women's magazines) as a real condition worthy of effective treatment options. *We were wrong, lady — it's not all just in your head; it's also in your ovaries.* (Actually, scientists had been researching PMS symptoms for decades and first labeled it as an actual "syndrome" in the 1950s.[7])

Though many hormones are involved in the cycle, let's focus on the two that ultimately impact outward behaviors more than any of the others — estrogen and progesterone. These two hormones control much of the cycle, but eventually their activity winds down as a woman approaches menstruation. After we explore the impacts of estrogen and progesterone, we'll look at what happens when their levels begin to recede during the premenstrual and menstrual parts of the cycle.

Estrogen: Curves and Competition

Estrogen is the Iron Lady of hormones, the fuel for the feminine engine. It is at its highest point during the first half of the cycle, the follicular phase, just prior to ovulation. But there is evidence that its mere presence at any level is related to physical attractiveness, sexual motivations, and competitiveness.

Women with higher levels of estrogen, particularly when estrogen peaks, are routinely rated by others as having more attractive facial features.[8] In one study, researchers used images of fifty-nine women who were photographed weekly for four to six weeks, and whose hormone levels were simultaneously monitored, to

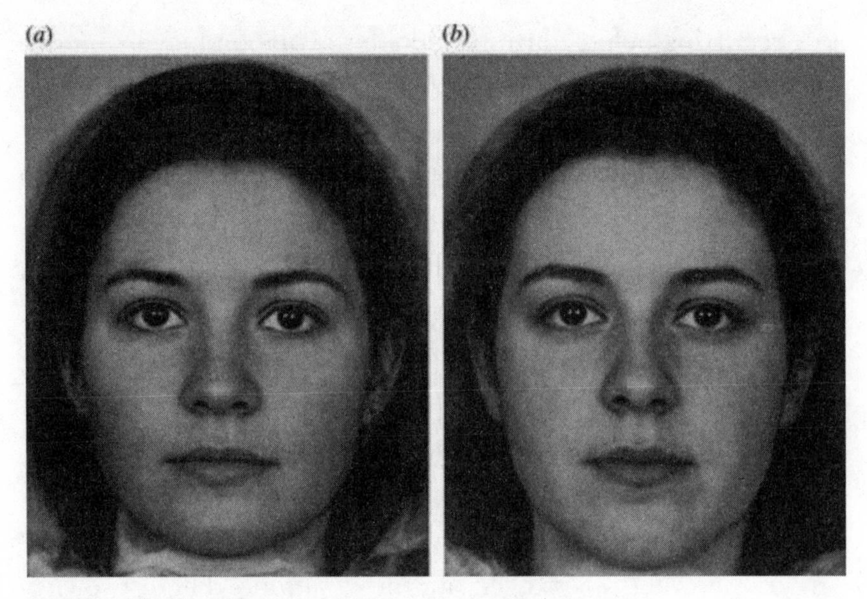

High-level (left) and low-level (right) estrogen composite

create two composite photos. Using digital overlays and mathematical formulas, they created a high-level estrogen photo and a low-level estrogen photo. When asked to select the face they found most appealing — specifically for "femininity, attractiveness, and health" — a separate group (composed of both women and men) selected the high-level estrogen photo.

Women with higher levels of estrogen also perceive themselves to be more attractive than those with lower levels.[9] (If you've seen the movie *Mean Girls*, in which the mirror is practically Best Supporting Actress, you've witnessed teenage estrogen gone wild.) Estrogen is why women grow breasts and accumulate more body fat in the hips and buttocks than men do. The classic hourglass figure, featuring large breasts and a small waist, is caused by high levels of estrogen, and scientists have shown that women with that particular body type have what they term "higher reproductive potential."[10] Assuming males are drawn to women with hourglass figures, the thinking goes, and factoring in that they have higher levels of estrogen, then chances of conception may be higher. (Of course, many men are attracted to more than one body type, and not all women with large breasts and narrow waists want to get pregnant or even succeed in getting pregnant. But the pertinent fact is that estrogen causes curves in the body, even if those curves are imperfect indicators of a woman's true fertility.)

These findings linking estrogen with attractiveness make sense. We know that surges in estrogen are responsible for much of what changes during puberty in girls. Their faces become more mature and womanly, as do their bodies. Onlookers perceive these estrogen-induced changes as attractive, likely because they indicate sexual maturity and possible fertility. Therefore, ancestral men who were indifferent to these features probably didn't reproduce as much as those men who did pay attention to them; likewise for ancestral women, sizing up the competition was important, even if the full-length mirror had yet to be invented.

Overall levels of estrogen are also related to the motivation to mate — and with whom. One study showed that women higher in estrogen preferred the faces of men higher in testosterone.[11] This is significant because it means that fertile women are potentially inclined to seek out males with good genetic material (or at least what might have been good genes ancestrally). (The stinky T-shirt study described in Chapter 2 is a cousin to these findings.) And high-estrogen women also report that they're more open to a sexual affair and feel somewhat less committed to their partners.[12] This could be because they perceive, perhaps correctly, that their mating options are better than those of other women and that they therefore have greater power to explore their options.

Higher estrogen is also linked to greater competitiveness[13] and possibly reduced fearfulness.[14] These findings could reflect changes that are associated with the cycle that make women more competitive for the best mating options — and that also happen to make women more competitive if they are generally higher in estrogen.

When estrogen is at its highest point in the cycle, and when women are at their most fertile, they move more — recall the pedometer study from Chapter 2, where women walked more at the midpoint of their cycles.[15] It's as if a kind of cabin fever sets in, with women needing to get out of the house and roam their surroundings when high estrogen kicks in. Interestingly, this uptick in physical activity parallels another behavioral change: A woman's sexual desires rise at high fertility but her calorie consumption drops.[16] Instead of going out to lunch, fertile women are going for long walks, possibly to check out their mating territory (I explore this more in Chapter 5).

Why would we exchange one behavior (eating) for another (roaming)? Is nature finally offering an effortless way to shed pounds? (*The Five-Day Estrogen Plan! Lose Weight While You Ovulate!*) Is there a strategic behavior at work here?

Women have a finite number of hours in the day. When fertile, we choose to be active and feel primed to mate; when not, we put up our feet, binge-watch Netflix, and eat some snacks. It appears that the motivation to mate when we're fertile trumps the motivation to feed our bodies — courtesy of high estrogen, perhaps combined with low progesterone. Such a shift in how we allocate our efforts is strategic, as I will show in the chapters to come.

So, high-estrogen women are more attractive, consider their mating options, are more competitive and less fearful, and prioritize some of their needs over others.

Imagine what fluctuations in estrogen do.

Progesterone: Protecting, Defending, and Playing for Both Sides

Progesterone is the loyal companion to estrogen, and it begins its rise and hits its highest concentrations during the second half, or luteal phase, of the cycle. The details are still unfolding, but one prominent theory is this: At a certain level, progesterone functions to suppress a woman's immunity, but paradoxically it can also reduce her risk for disease. These progesterone-driven processes are internal, but they still impact a woman's outward behaviors, as you'll see.

We all know our bodies are designed to fight off invaders — whether they take the form of a cold virus or the nasty microbes that ride in with a rusty nail. Assuming we're healthy, our immune system is programmed to go to war, calling up an army of antibodies to attack and defeat a potential infection. Therefore, a foreign body trying to burrow into the uterine wall would, under other circumstances, be recognized by the immune system as an enemy that needs to be neutralized. (Consider the body's potential rejection of a transplanted organ.)

But during the luteal phase, progesterone is capable of upending that reaction. When the uterine invader is actually a blastocyst (a fertilized egg that has already begun cell division), one that will eventually develop into a fetus, progesterone helps to prevent the immune system from attacking and allows the blastocyst to safely implant and settle into the womb.[17] Progesterone orders the army to stand down, so that human reproduction is possible.

Along with increased tolerance of the baby-to-be comes an increased risk of infection and a worsening of chronic infections.[18] But here's where progesterone really shows its upside as a double dealer: It causes a fortuitous *decrease* in diseases that are characterized by abnormally high immune response, such as too much inflammation.[19] (Normal inflammation is a key immune response, like the reddish, raised skin around a cut or the puffiness of a sprained ankle.)

Rheumatoid arthritis is an example of a disease in which an extreme inflammation response triggers painful joint swelling and eventual bone and cartilage damage. Pregnant women with rheumatoid arthritis report that their pain and symptoms lessen significantly. The hyperinflammation response seems to be suppressed by high levels of progesterone, which remain elevated throughout pregnancy, and which, in this case, have a healthful and protective effect on the mother, as her immune system is turned down to prevent rejection of a pregnancy. This trade-off between immune suppression and disease risk led my anthropology colleague at UCLA Dan Fessler to propose the "compensatory behavioral prophylaxis" hypothesis.[20]

According to the hypothesis, rising progesterone leads women (pregnant or not) to be especially wary of sources of disease — like germy people — given the immune-suppressing effects of progesterone required for a successful pregnancy. In one study, women in the high-progesterone phase of the cycle preferred composite facial photos of healthy individuals over those of unhealthy ones to a greater extent than women in low-progesterone phases.[21] (And, no, they weren't asked to choose between photos of Olympic athletes in their prime and photos of zombies on a bender.) The healthy faces, all shown here on the left, appear to be less grumpy and have clearer skin than those on the right.

Fessler's study[22] showed that women with higher levels of progesterone in their saliva were more revolted by disgusting images that connoted the transmission of disease (skin lesions, stained towels, parasitic worms, and even people riding in a crowded subway car). Furthermore, they were more likely to have

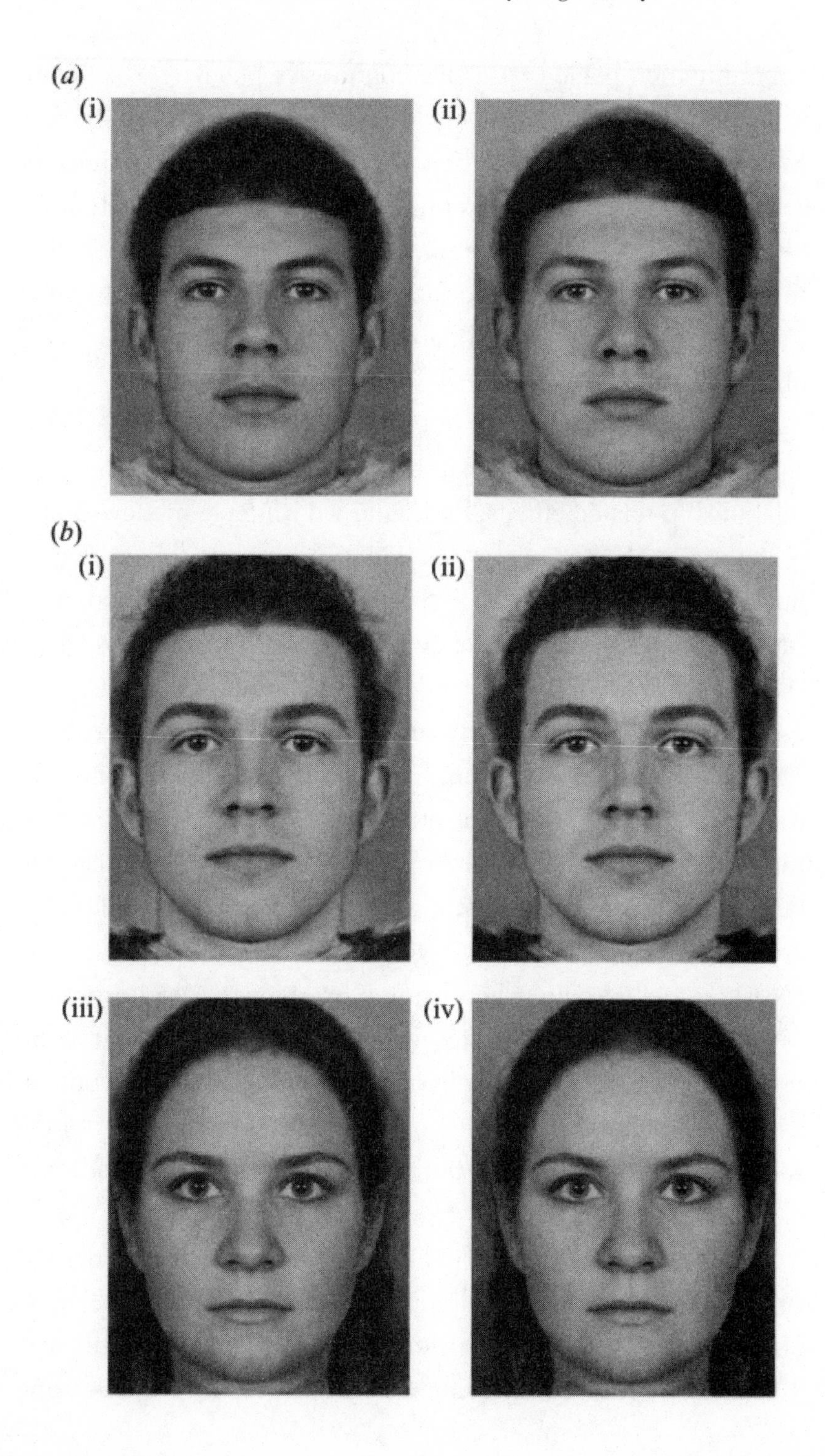
(a)
(i)
(ii)
(b)
(i)
(ii)
(iii)
(iv)

contamination-related obsessive-compulsive behaviors, like hand washing, if they felt they'd been exposed to germs. Similarly, they were more likely to pick at their skin (scabs) or eyes. Though this sounds like a behavior that would transmit disease, it's actually a behavior called ectoparasite removal, a form of personal grooming designed to eliminate surface parasites before they can enter the body — not unlike what baboons do for each other for hours on end.

Although the evidence is just beginning to emerge, and is somewhat controversial, progesterone also appears to have some relationship-related effects. Particularly during pregnancy, as progesterone continues to rise (as opposed to nonpregnant cycles, during which it drops precipitously), one might expect women to be particularly attuned to the quality of social relationships, given the help they would have needed in our evolutionary past (and today, too). Progesterone seems to be associated with faster categorizations of emotions on faces in computer-based emotion-recognition tasks and more attention to facial expression.[23] A high-progesterone woman can potentially read the room, weed out the frenemies, and make a good call about whom to lean on and whom to avoid.

On a related note, when women experience social exclusion (in a lab setting), progesterone has been shown to rise in response.[24] It could be rising in order to facilitate social reconnection. Progesterone is related to "affiliation motivation" among women[25] — or, more simply put, a desire to get together with others and be friendly.

Progesterone also can act as a mood mellower. In rodents and in humans, administering allopregnanolone, a neuroactive molecule that progesterone transforms into, produces calmness.[26] There's even preliminary evidence that progesterone quells some serious psychological maladies, such as suicidal ideation,[27] and that it improves the moods and symptoms of severe PMS sufferers.[28]

When you're "high" on progesterone, you are more grossed out by some people (and less tolerant of germy situations, like dirty public bathrooms), but you seek connections with others. Your immune system is suppressed, but you take steps to avoid the risk of illness. You also feel a little calmer (which is good if you have no choice but to use that rank restroom). Is this all true of progesterone? I would like to see much more research to know for sure, but I find these ideas intriguing.

PMS: Premenstrual Strategy, Not Syndrome

Estrogen begins its decline at the end of the follicular phase, right before ovulation, with progesterone starting its upward swing just as estrogen dips. Progesterone peaks midway through the luteal phase and then swoops downward, before menstruation begins.

It is during this end phase — the last days of the cycle, just prior to menstruation — that PMS can come calling. If progesterone is associated with feeling at ease and sociable, it's possible that the symptoms of PMS are actually a result of "withdrawal" from this hormone as it drops.

In the mid-1950s, British physician and researcher Dr. Katharina Dalton, who (along with her colleague Dr. Raymond Greene) coined the term "premenstrual syndrome," made a key connection between low progesterone and the onset of PMS.[29] Her ideas about hormones and behavior, which would impact women's healthcare, grew out of a personal observation she made about her own cycle: Dalton noticed that the migraine headaches that plagued her right before her period disappeared completely when she was pregnant; during pregnancy progesterone floods the system to support various aspects of fetal development and maternal health. And when progesterone bottoms out toward the end of the average cycle, it can impact female behavior in those PMS-y ways you're familiar with. Dalton is widely credited with elevating PMS to a legitimate medical condition with physical — not just psychological — symptoms. Though some have disagreed with her theories, she successfully treated numerous patients with progesterone in the world's first PMS clinic, which she established, and was quite interested in a link between postpartum depression and progesterone levels.

But back to PMS — the reasons it might exist in the first place may surprise you.

When puberty hits, girls (boys, too) ride waves of hormones they've never fully experienced before, and like first-time surfers, they wipe out regularly. And on top of that "normal" mood swinging, there is PMS. For me, PMS involved fighting with my sweet little sister, wanting to blast all the mean girls off the face of the earth, and torturing my patient mother. (When I told my mother I was studying hormones in my research, she smiled and said knowingly, "I'm not surprised." She knew that research is often "me-search.") I didn't feel calm, and I definitely didn't feel sociable. But, like most girls who grow into women, things got better as I learned what to expect with each cycle, and as I simply matured and gained more self-knowledge.

Once written off as a vague "female troubles" malady (in the same category with "hysteria"), PMS is very real. We've come a long way from those days when the condition was dismissed as just another example of "she's hormonal." Gynecologists and other experts in women's health can help with treatments, support, and resources.

By some estimates, about 85 percent of women will experience some form of PMS. Even with just one or two symptoms, it can be a bad batch of days: grumpy mood, occasional rage, skin problems, headaches, pelvic and breast pain, bloating, nausea, extreme thirst...there is a long list of documented psychological and physiological symptoms associated with PMS, as well as accompanying behaviors — many of which could be called antisocial.

The female behaviors associated with the most fertile part of the cycle, when estrogen is running high, are understandable from an evolutionary point of view: We know them as mate-seeking behaviors. But PMS behaviors seem to be the exact opposite: The symptoms associated with a drop in progesterone — like physical discomfort, antisocial feelings, irritability, and even a lack of sexual desire — seem to isolate women, as if something had misfired. But perhaps this behavior is actually strategic. PMS could be nature's way of driving a wedge between females and certain men, particularly when the reproductive agenda of the body seems thwarted by the onset of yet another menstrual period.

Women with PMS often find themselves particularly irritated by their spouses or steady boyfriends right before their periods arrive. *Just leave me alone.* That otherwise terrific and loyal boyfriend or husband is suddenly really annoying. *Would you please stop whistling?* Even if he is doing half the housework and/or childcare, he's still not up to par. *That is not how to load the dishwasher.* Our female ancestors never once had to remind their mates to put the toilet seat down, but they did have practical reasons to find certain men reproductively unworthy.

Here's the evolutionary logic: If an ancestral woman was having regular sex with the same male for several cycles without getting pregnant, then perhaps he was infertile or they were somehow genetically incompatible with each other. (Cases of infertility can be traced to women or to their male partners, or they remain mysterious — suggesting that perhaps some couples are just not compatible with each other.) After a few months of this, as her period approached and then arrived, it makes sense that she would eventually reject him and seek out other options. In modern times, a woman's mate doesn't get her pregnant every time they have sex (fortunately), so as her period approaches, this otherwise acceptable person might seem unacceptable. The antisocial behaviors associated with PMS may have evolved to ward off males who could not facilitate reproduction — guys with no game, or gametes.[30] (Why does PMS also affect our relationships with women? It could be spillover. That's my best guess for now, though I do not find it entirely satisfying. More grist for the research mill.)

While it may not make for the best of times, PMS could indeed have a purpose. And for those who slog through it monthly, there is some relief in knowing it lasts only a few days, if that, and the worst symptoms generally dissipate once a period arrives. Of course, cramps and blood aren't everyone's definition of "relief," but guess what? That phase of the cycle has its purpose, too.

Menstruation, Period

Though a menstrual period actually marks the beginning of the cycle, when hormone levels begin their gradual climb once again, we tend to think of it as an ending (and not just because it happens to be called a *period*). Perhaps this is because of how we tend to think of fertility, whether or not pregnancy is the goal.

If you have sex and are trying to get pregnant, then there's some suspense as to whether or not you'll get a period when you're due for one. If it arrives and you're disappointed, it does feel like an ending of sorts, though you can try again soon. On the other hand, if you would like to avoid pregnancy but had a precautions malfunction, you're relieved when you get it, and that also feels like some form of resolution — as in *Whew... glad that's over with.* So, though medical science considers Day 1 of a menstrual period to be the beginning of a new cycle, it really does seem like an ending. (Plus, who wants to mark a fresh beginning with a spiffy new... box of tampons?)

Unlike other phases of the cycle, when major hormones soar or plummet, during a menstrual period they are fairly steady, at a low level. Periods last anywhere from three to five days and can be as short as two and as long as seven days — a reminder that not everyone has that "average" twenty-eight-day cycle. A period length can vary based on many factors, including age. A fourteen-year-old girl may have a much longer and more erratic cycle than a woman in her twenties or thirties, since cycle length starts to shorten and become more regular as a woman matures (and then becomes irregular again as menopause approaches).

This variability in length probably doesn't get as much attention as it deserves, and that may explain why some women encounter fertility problems. If women have healthy cycles that are twenty-one days or thirty-five days or even longer, they don't fit into the twenty-eight-day template that so much mass-marketed fertility advice is based upon. A fertile woman may not be ovulating on Day 14 — she may be ovulating on Day 10 or Day 18. And if that's the case, all baby-making bets are off (as are "natural" birth control methods like the rhythm method). This is why knowing the length of one's own cycle, and expecting some variations, is so important. We'll talk more about fertility and birth control — and the hormonal misunderstandings that can derail both — in Chapter 7, particularly the box "How to Succeed in Baby Making without Really Trying."

Free the Tampons!

There are costs and benefits associated with estrus, a human equation that is generally balanced like most things in evolution—but the costs column is decidedly lopsided when it comes to one feminine fact: the price of a period. If a modern woman has, on average, four hundred periods, that is thousands of dollars spent on pads and tampons alone. The financial burden is very real for some households, and in large cities such as New York, schools are starting to offer free supplies in middle and public school girls' bathrooms—a move championed by the Free the Tampons Foundation (freethetampons.org), which believes that tampons and pads should be free in every public restroom, just like soap and toilet paper.

In a nod to a cost that women alone bear, there is a move afoot to drop the "tampon tax" in some states. But "feminine hygiene products" aren't the only area that makes me see red, so to speak. More women—and men—are increasingly motivated to tackle the so-called pink tax, a reference to the fact that many consumer products—toys, clothing, soap, shampoo, and more—for girls and women are more expensive than versions of the same products for males. (One consumer group notes that a girl's scooter at a famous big-box chain costs more than a boy's—the only difference being that the girl's model is pink![31]) During the legislative debates on American healthcare, one (male) lawmaker questioned why prenatal care was included in mandated health insurance—why should men have to pay? (Perhaps because they'd like their mates and offspring to be healthy so that we don't hit an evolutionary dead end?)

Women have been having periods for hundreds of thousands of years, and we've come a long way from the days of using rags, as women did into the twentieth century. Tampons and pads are far more convenient, but women—86 percent of whom report getting their period in public without having supplies on hand[32]—shouldn't have to pay a premium for them. A more economical reusable menstrual cup has been gaining in popularity (not for everyone, granted, but reusable cups are environmentally friendlier than tampons with plastic applicators, which wash up on beaches the world over). But when you are bleeding and banging on a broken vending machine for a seventy-five-cent tampon, you have to wonder when we'll get better at the business of menstruation.

We may have to wait for free tampons in every public women's room, but the elimination of the sales tax on feminine supplies is beginning to happen. If it hasn't happened in your state, perhaps you'll be moved to write a letter or engage in your own period-power activism—and while you're at it, ask toy makers why pink plastic costs more than blue.

Despite the caricature of the pale (because she's losing blood) woman retreating to her bed with a heating pad, sitting out gym class, "ragging" on her put-upon partner, or causing bloodthirsty sharks to beach themselves, there is not much outward behavior during menstruation that may be linked to hormones. Remember, the so-called raging hormones have quieted for now, and the emotional fireworks, if there were any, were lit up and set off during PMS. Certainly, cramps, headaches, and other symptoms can be irritating. (Severe cramping and heavy bleeding are not normal and can signal conditions such as fibroids or endometriosis.) But with the exception of maybe some extra bathroom breaks, a normal menstrual period is generally business as usual for most women. It's just life.

So, what's the point of a period, anyway? You now know that the bleeding occurs because thickened endometrial tissue along the uterine walls is being shed since it's not needed to support a blastocyst. Humans, other primates, elephant shrews, and some species of bats are among those mammals that have menstrual periods. But in other mammals, there is no built-up endometrium discharged through menstruation; instead, the uterine lining and tissues thicken with blood and other nutrients only if fertilization occurs.

One theory about menstruation, largely dismissed but still discussed, is that female bleeding serves to flush out "bad" sperm that may carry bacteria, viruses, and other pathogens. This idea was famously put forth by the controversial evolutionary biologist Margie Profet, who theorized that menstruation — a biologically inefficient process given the regular depletion of blood and tissue — was nature's way of removing disease-causing agents from the female reproductive system. "Sperm are vectors of disease," she wrote. (Profet would later theorize that morning sickness was also an evolutionary response to foods potentially toxic to pregnant women — throwing up or being repulsed by certain food odors was nature's way of reducing exposure to dangerous allergens, risky germs, and even carcinogens that could harm mother and baby.)[33]

But sexually transmitted diseases like chlamydia still prevail; we might think that STDs would be harder to transmit if they were mitigated by the average four hundred periods a modern woman has in her lifetime. (Actually, Profet attributed this fact to the use of oral contraception pills that reduce or even suppress menstruation.)

Finally, if getting rid of germy microbes was the reason for menstruation, why did only a handful of species evolve to get rid of disease-causing agents? Are all the rest doomed? (And will *Planet of the Apes* be supplanted one day by *Planet of the Elephant Shrews*?)

There is another theory regarding menstruation, but it has less to do with the bleeding itself and more to do with reasons for the thickening of the endometrium. Even when a woman is not pregnant — even if she doesn't have sex during her most fertile phase — the endometrium still plumps up to prepare for the implantation of a fertilized egg. But as mentioned previously, in a nonprimate animal the tissues do not thicken *unless* pregnancy has occurred. Unlike other animals, we're going to a lot of trouble for a guest who may never arrive!

Researchers have surmised that perhaps the thickening process is linked to the particularly aggressive nature of the human embryo, which implants deeply in the uterine wall in order to fully access life-supporting maternal resources (including blood vessels).[34] (By contrast, other species develop superficial placentas that don't "invade" so deeply.) Here is where the concept of "maternal-fetal conflict" comes into play: In this scenario, the human embryo is literally digging in and hanging on for dear life;

at the same time, the mother needs to protect her own life (for her own sake as well as for that of future offspring). So, in preparation for this burrowing guest, her body responds through a hormone-triggered balance of self-defense and accommodation. High progesterone levels cause the thickening of the endometrium in case this tiny interloper arrives. And if he or she does not show up, progesterone levels fall and the lining is eventually shed.

A related idea is that it's too metabolically expensive for the body to maintain the particularly large human endometrial lining of the uterus — so it is shed every month, rather than staying in its ready-for-baby state. In fact, University of Michigan anthropologist Beverly Strassmann estimates that there is a 7 percent lower metabolic rate during the preovulatory phase than during post-ovulation, when the body is preparing for the possibility of implantation. She estimates that over the course of four cycles, women could conserve as much as six days' worth of calorie expenditure by ridding themselves of the uterine lining through menstruation.[35] That could have made a huge difference to our human female ancestors living close to the margins in the availability of food.

At the end of the day (or rather, the cycle), there does seem to be evidence that the thickening of the endometrium, and the subsequent shedding and bleeding, is an adaptive response to the aggressive quality of the human embryo (and those of a select group of other mammals). Biologically, it seems very inefficient to build up blood and tissue, only to lose it all on a regular basis. And yet, perhaps our hormones — despite their cyclical gifts of blood, sweat, and tears — are doing something very efficient: They're protecting our own survival and that of our future offspring.

4

The Evolution of Desire

We know for certain what estrus looks like among animals in the lab (a certain four-legged female, running on her wheel) and in the wild (her uncaged, free-range cousin, wiggling her ears and hopping about to get attention). We've also explored what appears to be estrous behavior in humans — and now know there's a lot more to it than stinky-in-a-good-way T-shirts. However, it's not as straightforward as women suddenly becoming interested in having lots of sexual intercourse on those all-important high-fertility days of the cycle. Recall that the evidence is mixed, at best, regarding a woman's desire for frequent sex at high fertility.

Perhaps, as the stinky T-shirt study hints, just as mice and dogs in estrus strategically select particular mates, for women, *not just any male will do.* If that is indeed the case, women will choose certain types of men. They will be drawn to some potential partners and at the same time put off by others. And they will seek out specific traits when they are most fertile within their cycles, and perhaps have different preferences when they are outside of the fertile window. The research I will share with you shortly resoundingly supports these predictions.

These hormonally triggered shifts in sexual behavior across the cycle are fascinating and complex and at the heart of much of my research. I believe a woman's sexual behaviors — her desires as well as her actions — serve clear purposes that can define her destiny, as well as the destiny of her potential offspring.

But why did we evolve to have estrous behaviors in the first place? As I've mentioned, some humans (male and female) are uncomfortable acknowledging how much we have in common with animals, especially when it comes to sex and mating. Yet, women do possess uniquely human behaviors, notably the desire and ability to have sex outside the fertile phase for reasons other than reproduction. (And of course, men share this desire and ability.) In that way, through their far less constrained sexuality, humans took a transformative detour away from much of the animal kingdom.

This was sexual behavior driven by hormones, and as you will see, it most likely evolved because of our increasingly big brains and the needy human offspring that resulted, dependent children who fared best when they received care from both moms and dads.

Lizards in Heat

Vive la différence, but as evolution tells us, it wasn't always that way.

Nearly half a billion years ago, female and male vertebrates diverged in their reproductive machinery, and each developed their own types of estrogens and corresponding hormone receptors. (If you haven't thought about "hormone receptors" since ninth-grade biology, think of them the way biology teachers have been explaining them for generations. Hormones are like tiny messengers traveling to specific cells in the body. Hormone receptors are the intended receivers. If the message reaches its recipient, then the cell will react accordingly. In the case of sex hormones, they will turn on genes within the cell that can cause the body and brain to change — to grow reproductive tissues and direct our desires, for example.)

The hormones and receptors eventually evolved to function differently in male and female brains and bodies. In females, estrogen triggered egg maturation — and likely led to changes in sexual motivation to facilitate reproduction. In turn, males evolved to detect any external signs of estrogen, such as changes in female scent, and to find them especially sexually attractive (as we'll see in Chapter 6).

All vertebrates, from Gila monsters to mice to chimpanzees to humans, share behavioral estrus and the changes in sexuality that accompany shifts in estrogen. The branches of the evolutionary tree that define the genetic relationship among species and how they split off into new species over deep time give us major clues about the point of origin of those features, including estrus, that species share with one another (see the illustration of the evolutionary tree on the next page). We know from the tree that this ancient hormonal dance existed before lizards and mammals diverged.

Approximately four hundred million years ago, the estrogen receptor became sexually dimorphic (different between males and females), and this was the likely origin of estrus. The top species (from monotremes to teleost fish) are those presumed to have estrus, including dinosaurs! There was, of course, further divergence between species at the end of each branch. *Homo* (humans) are broadly lumped in with all animals with placentas. However, the human version of estrus, though it does bear similarities to

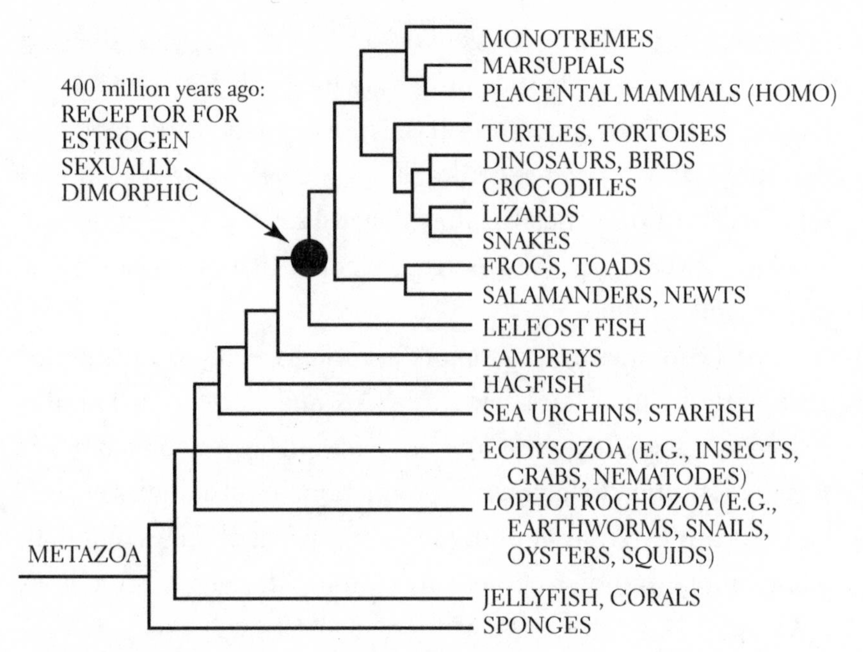

The evolutionary tree showing genetic relationships among species

that of the animals on its branch of the evolutionary tree, has its own unique features.

We evolved from what appear to be stunningly different life-forms. Amazingly, we owe a lot to our fishy ancestors, who were the first to be "feminine" — in the sense that their estrogen receptors worked differently from those of males, creating a distinctive female reproductive type, controlled by estrogen. It took millions of years, but eventually that same estrous cycle would take hold inside of our earliest human ancestors, who would lay the foundation for strategic sexual behaviors among females.

Let's look more closely, then, at the "evolution of desire," as evolutionary psychologist (and my PhD mentor) David Buss so aptly phrased it in his classic book of the same title.[1] If we can elaborate on our understanding of the point of changes across the cycle in humans and what, specifically, they accomplish for women, perhaps we'll be able to answer many questions about human sexual behavior. Investigating the origins of estrus can also help us understand why some relationships can span a lifetime while others flame out in a weekend.[2]

The most obvious explanation for human estrus, as discussed earlier, was that it provided a pathway for ancestral females to chase after something that meant the difference between sending offspring into the future and disappearing from earth without a trace. That something was sperm.

But if the only driver during estrus was some kind of general hunt for viable sperm from *any* male to fertilize an egg, if absolutely *no* discretion was exercised by females when it came to mating for reproduction, take a moment to imagine the consequences, both genetically and otherwise. A girls-gone-wild scenario doesn't make a lot of sense, particularly when you consider the parental investment theory put forth by evolutionary biologist Robert Trivers (discussed in Chapter 2): Among the sexes, the one that has to invest the most in reproduction will be more selective in choosing a mate. Females are limited in the number of offspring they can produce, and once the child is born, mothers invest hugely in their care. Therefore, it's unlikely they would have made hurried mating choices — just to obtain sperm. ("Baby fever" as we know it probably did not originate with ancestral females.)

Furthermore, when it comes to mating, there is no evolutionary reason for females to do all the work (they'll expend plenty of energy once they become pregnant and bear offspring). In the wild and across many species, males are very capable of seeking out and finding fertile females, in part because estrous females aren't exactly hiding their light under a bushel — consider the sexual swellings of female baboons. Even female fish emit pheromones to signal their availability for mating.

There is another reason why the sperm-or-bust explanation is flawed. Scientists failed in their initial searches for evidence of human estrus because they were looking for love in all the wrong places. Based on what they'd observed in animals, these early heat seekers predicted a *general* increase in sexual activity among women at high fertility. Frenzied females, remember? But because these scientists were operating on outdated notions of "heat," and since there was no research to demonstrate otherwise, they couldn't take into account the fact that fertile females are, as it turns out, very selective about potential mates.

In the decades that followed the T-shirt study and other early work, we saw increasing research that demonstrated that a woman's sexual desires change across the cycle, including the type of male she finds most attractive when her fertility is at its peak. Furthermore, a woman's preferences aren't merely for certain physical features (for instance, more symmetrical features that possibly indicate genes with few flaws, as described in Chapter 2), but for specific behavioral qualities as well. A number of studies have demonstrated that during their fertile phase women prefer men who are more confident and dominant (even arrogant), behavioral traits that track with this idea of a more "masculine" male.[3] The same preferences have been observed in females of other species, including primates, who similarly choose high-ranking males to mate with. Even in Gila monsters, males engage in a scripted dominance ritual, and the dominant male wins the reproductive prize, the estrous female.

If only the he-men were being selected, this meant that a lot of other males were potentially left by the wayside, including those candidates who may not have had the biggest biceps but who made up for those supposed deficiencies in other areas, as we'll see. While the Sexy Cad offered something that females wanted at least some of the time (during estrus), the Good Dad could offer other qualities that females — and their offspring — needed all of the time. But before we determine why (and if) nice guys seemingly finished last, let's look at why fertile females targeted Mr. Sexy in the first place.

Bad Boys and Good Genes

Imagine an ovulating female in search of a mate to father her offspring. Numerous willing and capable males are standing by, but only one — the dominant male — will do. He's aggressive and confident, loud, a bit egotistical, and physically larger than the others. But he's "the one." Whether we're talking about wolves, birds, nonhuman primates, or humans, this is the mate-selection behavior that emerges across the species over and over again: females at high fertility showing a preference for dominant males. But that's on the surface — as there is much more to it than the flash of a peacock's tail, the loudest bark or howl, or rock-hard abs. And of course, one male — even if his floppy flanges are huge — can't necessarily father all the offspring.

We know that the landmark stinky T-shirt study by psychologist Steve Gangestad and biologist Randy Thornhill showed that women at high fertility favored symmetrical features in potential mates; by selecting these prime specimens, women were also choosing candidates with beneficial genetic material that they could pass on to their offspring. In ancestral women, much was riding on their mate selection, including the viability and survival of offspring. That, in a nutshell, is the good-genes theory of mate selection: When conception is possible, females show a preference for males with traits associated with fit genes they can pass on to their offspring, as demonstrated by their dominant behaviors and certain physical attributes, because both females and their offspring will ultimately benefit.

The good-genes theory bolsters the idea of strategic female sexuality, offering evidence for why females evolved to be picky about whom they might choose. But human estrus accounts for only a narrow window of time within the hormonal cycle, about five days — up to and including the day of ovulation. Furthermore, a woman at her most fertile may *prefer* a good-genes guy, but that doesn't mean she'll necessarily act on her desire. She may not get the opportunity to act within the fleeting fertile period, or she may choose not to act on her attraction — particularly if she's coming to realize that, in the long run, she'd be better off with someone else (we'll get to that part in a moment).

Among all males capable of contributing to the reproduction of offspring (that is to say, providing sperm), only a handful of them will qualify as the textbook good-genes fellow — logically, not every male in a group can be dominant, or else there would be no . . . domain. There are a lot of other (nonalpha male) fish in the sea, so to speak, and on top of that, there are a lot of other days in the cycle to account for besides estrus, and the males most preferred on those days could be offering other benefits besides their genes.

Cues of good genes may have had huge consequences ancestrally, and many animals, including nonhuman primates, still prioritize their mate selection during peak fertility based on the biggest and the best. But human females, in keeping with their shifts in desire across the hormonal cycle, eventually added a new sexual strategy on top of the estrous behaviors they inherited from the long line of animal ancestors: long-term relationships. Through the evolution of pair-bonding (the selection of a single partner) and extended sexuality (having sex outside of the fertile period, which humans do *a lot*), we took a human turn.

Pair-Bonding: It Takes a Village (and a Partner)

The demands on ancestral women during and after pregnancy were particularly huge — pregnancy, birth, lactation, child-rearing, and the likelihood of other dependent children. On top of that, compared to nonhuman primates, it took much longer for a large-brained human child to reach certain developmental milestones, such as walking and self-feeding, that would contribute to his or her eventual independence from a family group and survival. Chimpanzees are entirely self-sufficient by the age of four. In contrast, human offspring are heavily dependent on their parents for many years and in most traditional societies cannot gather enough food to feed themselves until the age of twelve — chips and Halloween candy excluded; we're talking about food for survival, not snacking. (When I ask my students how long they think human children are dependent upon their parents, they get all shifty-eyed and uncomfortable in their seats, perhaps contemplating their own student loan debt, or the need for a place to crash after graduation — with health insurance. We usually decide it's at least age thirty.)[4]

Humans evolved to have an "expensive brain" framework — meaning that our complex, large, and slower-growing brains have high-quality caloric and nutritional requirements, relative to other mammals.[5] Or, to put it another way, we're smarter than other animals, but it comes at a price — high levels of premium fuel for cognitive development. Among ancestral humans, meeting that cost — through consuming and absorbing enough calories, or through conserving physical energy to compensate for caloric expenditure — was truly a matter of life or death.

Nonhuman primates hit the "gray ceiling" — as in gray matter, meaning that their brains developed only up to a point and then they stopped — but we humans kept going, and access to nutritionally dense calories was part of the reason. We gained one serious edge when we eventually figured out ways to extract more calories and nutrients from our food supply, and that "brain food" helped us shatter the gray ceiling. That food supply consisted of foods that were gathered — mainly plants — and foods that were hunted, like big game.

We'll never know for sure whether it was a man or a woman who first learned to use tools to cut and fire to cook meat and other foods, probably increasing their nutritional density. Instead of gnawing on tough animal flesh, we learned to tenderize it through pounding and cooking. Instead of passing over vegetables or fruits with hard shells or impenetrable skins, we could slice them open. Now we could chew and digest more animal protein and access other macronutrients that would contribute to brain growth and cognitive development.[6]

But it wasn't just a matter of consuming more (and better) calories, thanks to the world's first home cooks; it was also a matter of expending fewer calories by redirecting energy expenditure. For instance, primates dedicated vast amounts of energy to climbing up and down trees and traveling long distances for food and shelter. Humans, moving well on two feet, developed less demanding habits — though life was far from easy, particularly for childbearing women.

Clearly, it was especially difficult to be a single parent if you were an ancestral female who had to invest significant resources to bring forth and raise a big-brained child — physical energy, nutrients, and time that stretched into years. Unlike chimps, human children were tied to their mothers well beyond age three or four, completely dependent on maternal support for their survival. These days, a lot of twelve-year-olds mark their "independence" with their first smartphone, but back then, there was no app for hunting and gathering or using tools.

Someone had to step up and help to protect and provide. It's possible that a mother's extended family would have assisted in raising her dependent children, but a father would have been the most closely related potential helper. Fathers, as the nearest relatives of these children, faced more evolutionary pressure — by virtue of the shared fate of their genes — to invest in assisting with their offspring than anyone else in a mother's family group. So, men evolved strong inclinations to invest in their offspring (though some do adopt the alternative love-'em-and-leave-'em strategy). "Co-parenting" has such a modern ring to it, but it has its roots in our earliest human history.

And here is where humans start to become more . . . *human*. During estrus, ancestral women attracted and were attracted to dominant, alpha males who could provide good genes; but outside of that narrow window of fertility, females developed strategies to attract and stay bonded to males who would invest valuable resources in them and in their offspring for the long term — males who would offer their time, provide protection, secure food and shelter, and more. Scientists use the phrase *pair-bonding* to describe this steady relationship arrangement.

Or think of it this way: The Sexy Cad was beginning to face serious competition from the Good Dad.

From Brawny to Brainy

We humans have big brains compared to nonhuman primates. A full-grown chimpanzee's brain, for instance, weighs about four hundred grams; a mature human brain weighs more than three times that. But at birth, the tables are turned.

Our primate relatives are born with comparatively large brains that grow rapidly during gestation. In contrast, our human brains experience extreme growth *outside* the womb. The human birth canal is capable of great expansion during the birthing process, but it's still too small to allow for the safe passage of an extra-large skull. (One theory about the human birth canal's limited capacity is that as humans evolved to become bipedal, rather than continuing to climb around on all fours, the human pelvis narrowed.) Therefore, full-term human babies essentially have premature brains, but those brains will grow rapidly over the next few years.

By age two, the human brain is physically at about 80 percent of its adult size, but in terms of cognitive change and development it doesn't reach maturity until the mid-twenties.[7] The final product is a complex, large brain that may be, on average, only 2 percent of our body weight but requires 20 percent of our oxygen and blood supply, as well as significant energy (calories from food) to function properly.[8]

Though ancestral women were probably the original supermoms, it would have been impossible to sustain successful reproduction over the generations without assistance from extended family and, ultimately, from fathers. The complex, nuanced functions of the human brain owe more than a little to the rise of pair-bonding. Without the investment of a father—a helper at the nest—ancestral women would have had great difficulty providing enough calories and other resources for developing offspring, and without enough brain-boosting nutrition, humans would have hit the "gray ceiling," as nonhuman primates did.

Extended Sexuality: How Pairs Stayed Bonded

The rise of pair-bonding does not mean that ancestral women completely threw over the sexy-but-not-always-reliable alpha males for their less dominant (but more dependable) wingmen. But it does suggest that females developed strategies outside of estrous behaviors to attract males who would invest in their offspring. Practically speaking, there were simply too few alpha males to help with child-rearing — demand exceeded supply. Furthermore, alpha males hardly needed to offer their help. They were being chosen as mates for their "good genes," not for their willingness to stick around and share the care. But there were plenty of other males, symmetry be damned, who had much to offer in the way of time, protection, food, and other resources.

One way to ensure that these less sexy but stable males would be there for the long haul was for females to be receptive to courtship and sexual intercourse outside of the narrow window of fertility, at any point in the cycle — and not just receptive, but proactive, as initiators. Having sex for reasons other than reproduction — *extended sexuality* — is a modern human behavior that had its roots in the need for help at the nest.

Mammals like cats and dogs, for instance, don't copulate outside of estrus. (And for the most part, unless it's *The Aristocats* or

Lady and the Tramp, male felines and canines aren't known for remaining forever loyal to the mothers of their litters and providing paternal assistance.) However, some primates such as orangutans and chimps do have intercourse during nonfertile phases of the cycle, and bonobos are famous for using frequent sex to make peace, not just little bonobos.

One theory on why these primates engage in nonreproductive sex is that it creates "paternity confusion" and therefore can reduce male aggression toward offspring. If a female allows copulation with multiple males, even outside of the fertile window, those males may think that they're the father and won't harm or kill her offspring.

For humans, however, the reasons for extended sexuality seem to be entwined with pair-bonding. Females in search of mates, it turns out, have two sets of priorities:

1. Mr. Sexy (aka the Sexy Cad): the male she finds sexually appealing, who satisfies her preference in a more short-term way. She is most attracted to him during the handful of peak fertility days — when her good-genes radar is tuned on high.

2. Mr. Stable (aka the Good Dad): the male who is not as sexually attractive during peak fertility but who is otherwise appealing as a kind and caring mate whom she'd like to be with over the long term — in part because he is more likely to stick around and help with offspring.

Some women will be able to attract the rare man who combines the best of both. (This paradigm of the devilishly handsome fellow who is also an excellent protector and provider is at the core of many a romance, from Lizzie Bennet's Mr. Darcy to Carrie Bradshaw's Mr. Big — and even Edward the vampire from *Twilight* and Christian Grey from *Fifty Shades of Grey*, though envisioning these guys as having been tamed into truly good dads seems like a big stretch.) Such females, satisfied with their two-in-one mates, are unlikely to look around for other options during estrus, or perhaps at any time across the cycle.[9]

However, for the majority of females who don't land on this ideal combo (and that is most women, as there's that real-world problem of supply and demand), nonalpha males will start to look better and better. Later on, I'll share with you more studies that document these shifting preferences across the cycle.

Because the theory of extended sexuality states that females are willing to engage in sex across the cycle — including nonfertile days, weeks, and years, throughout all phases of life from youth into old age — some scientists interpret this to mean that human females do *not* have estrus, specifically "classic" estrus (with reproductive sexual behaviors limited to fertile days only). In other words, the thinking goes, if we like to have sex when we're not fertile, then we can't possibly be experiencing estrus, right?

As you know, I disagree. I think a better way to think about extended sexuality isn't that it rules out estrus, but that it is a complement to estrus, as argued by Nick Grebe and others.[10] This strategy offers women another way to shore up pair-bonded male investment in relationships and offspring. Women's receptiveness to sex also makes sense from the male perspective. If ovulation is concealed, it's better to try without the "success" of reproduction than to miss a crucial reproductive opportunity (from an evolutionary perspective, it's better to be safe than sorry).[11]

Extended sexuality with a pair-bonded partner also accomplishes something else for women. It allows females to gradually shape their relationships with males more carefully, selecting them not only for their good genes, but for their good behavior as well. As we're about to see, across the cycle a woman's preferences change depending upon what she is looking for: a long-term mate or a short-term partner. These cycle shifts allow women to come closer to combining the best qualities of Mr. Sexy and Mr. Stable.

It took a while (half a billion years), but "the one" was finally on the horizon.

Beetle-mania

Extended sexuality isn't just for mammals. Even insects, like carrion beetles, exhibit extended sexuality, copulating with their pair-bonded partner (as well as with other males) outside of their fertile period.[12]

Researchers have observed that female "burying" beetles go even further when it comes to getting males to stick around; when her offspring are at the delicate larval stage, the female is able to produce a hormone that not only blocks her own egg production, but also serves as an "anti-aphrodisiac" to her partner to get him to shift away from mating and toward caring for offspring. The male otherwise can't get enough copulation—perhaps because he wants to ensure paternity. Males have even been observed mounting females while they are in the process of laying eggs. They aren't being selfishly horny; they're most likely concerned that other males, drawn by the scent of the carcass the beetles inhabit and the presence of a fertile female, will home in on their partner, so they keep on doing their duty, to ensure that no other males become the fathers of the offspring that are to come. Males, as we know, have their own strategic behaviors.

When the beetle sex stops, the childcare begins, since both parents work to feed the larvae. Priorities have shifted! Once baby beetles are able to crawl on their own six legs and scuttle out of the nest (in this case, a rotting carcass), the adults are freed up to resume relations without being interrupted by the need to care for the little ones. In this regard, as biologist and beetle researcher Sandra Steiger says, "they are a very modern family."

Why Estrus Desires Evolved

At the heart of how you choose a mate is this seemingly impossible question: *What are you looking for?* The answer can be complex or it can be to the point. It depends on whom you are asking, and when. But these answers are incredibly revealing, and they may contribute to our understanding of the behavioral shifts women experience across the cycle.

In other words, you're not hormonal. You're making important decisions.

One-Night Stand versus Golden Wedding Anniversary

When you ask men and women what qualities they most desire in a long-term mate, they will prioritize the same characteristics: kindness, intelligence, and a good personality. This is what David Buss discovered when he surveyed thirty-seven cultures on six continents and five islands around the globe, in a landmark study published in 1989.[13] The answers from both sexes — more than ten thousand strong — were remarkably consistent and seem to reflect the response you would get today if you were simply chatting with a friend. Someone who is nice, smart, and great to be around — what's not to like? You may kiss the bride, or the groom.

But that's where the happy ending ends, or at least gets more complicated. Here is the list of top-ranking qualities men and women value in marriage partners (long-term mates), from Buss's study:

Men

1. Kindness and understanding
2. Intelligence
3. Physical attractiveness
4. Exciting personality
5. Good health
6. Adaptability
7. Creativity
8. Desire for children

Women

1. Kindness and understanding
2. Intelligence
3. Good health
4. Exciting personality
5. Adaptability
6. Physical attractiveness
7. Creativity
8. Earning capacity

The similarities are striking, but the differences are also notable. Men value physical attractiveness in a long-term partner more than women do. Women want healthy men who can earn a living. Women do value attractiveness — it comes in at number six. So, they might very well be seeking the Mr. Sexy and Mr. Stable combo that makes for a good protector and provider. Of course, this list doesn't tell the whole story of marital relationships. But it's an intriguing look not only at "what women want" but at what men want, too.

The picture changes when you reframe the question and ask men and women what they're looking for in a short-term partner, which is precisely what Doug Kenrick and colleagues did in their research.[14] They asked men and women what the *minimum acceptable level* would be for someone they would have a dating relationship with, have a single sexual encounter with, marry, and so on. For men, physical attractiveness for a onetime sexual encounter was *not* as important as it was for steady dating or marriage. For women the pattern differed — physical attractiveness was not as crucial for steady dating or marriage as it was for the onetime sexual encounter. In fact, the minimum level of attractiveness women required for a sexual encounter was greater than the level men said they would require.

So, for women, physical attractiveness was very important for a one-night stand. This preference seems consistent with the good-genes theory. To put it plainly, for dinner and a movie, a guy doesn't need to be all that good-looking. But if sex is involved in a onetime hookup, then he'd better be ridiculously handsome (and very symmetrical), because he'd only be offering genetic material. (Of course, this is not necessarily how it plays out on Friday nights all over the world; rather, it's an attempt to understand why we like what we like.)

Short-term preferences aside, if men want to marry physically attractive women who will give them children, and if women want to marry healthy men who can hold down jobs, you may be wondering how it's possible that we humans have lasted on earth this long. How did we get together and overcome our different preferences, particularly our contrasting prioritization of physical attractiveness? Some may think the answer lies in the things men and women ultimately have in common — the kindness and understanding, intelligence, and good personality both sexes seek out and value, first and foremost, in a long-term mate. Sure, cue the violins, but... before we got to "sunrises, handholding, and long walks on the beach," we had to crawl out of the swamp, and females had to square their estrus desires with the reality of supply and demand.

The Evolution of Trade-Offs

It's unclear who first strung together the words "tall, dark, and handsome," but women have added to that list of qualities when asked what they're looking for in a man. *Smart, funny, curious, cooks and cleans, compassionate, handy, bird-watcher, likes crossword puzzles, strong, movie buff, bookish, sporty, ripped abs a plus (but not too ripped), calls his mother*... You get the idea — the wish list varies from person to person, and it tends to be a long one.

In the ideal world, we'd find the ideal mate. But we live in the real world, so we look at our real options. And that's why females have learned to make trade-offs when they search for a long-term mate, opting for the males with whom they'll be most compatible over time.

As we know, not all women can be successful in finding that perfect combination of Mr. Sexy and Mr. Stable. (There never was and never will be enough well-behaved symmetry to go around.) But based on the theorizing from my lab and others who have investigated these dynamics, here's what might have happened when an ancestral woman *did* land "that guy":

She maintained that valuable relationship by combining two sexual strategies: estrous lust for his high-fitness genes and a nonestrous attachment to his good-partner traits. During ovulation, she secured the genetic material she wanted for her offspring from his Mr. Sexy side; during her nonfertile period, through pair-bonding and extended sexuality, she established a nonestrous attachment to his Mr. Stable side. And most of the time, she was not in estrus, so this latter attachment was strategic indeed.

Good for her, but what about the rest of the women, as in most of the female population? In their search for long-term male partners, given the exceedingly short supply and high demand for alpha males with a sweet and caring side, it seems that women learned to make trade-offs. They exchanged sexy for stable, picking the reliable male who would help at the nest (pair-bonding), and maintaining that important relationship by being sexually available (extended sexuality).

Still, there were bumps in the relationship. Even if an ancestral woman partnered with the Good Dad because of his ability to provide help with offspring and material resources, her estrous preferences would still emerge during her peak fertility, meaning she'd be drawn to the Sexy Cad. At this point, she had two choices: act on her desires during ovulation and discreetly secure Mr. Sexy's high-fitness genes, or not. Doing so was an extremely risky strategy, particularly for women who already had children. (In addition to being violent toward their mates, jealous males, even Good Dads, were capable of harming or killing offspring who they discovered weren't their own.)[15]

So she made a highly practical trade-off to ensure the survival of her offspring and of herself. She stuck with Mr. Stable, whose qualities of cooperation, reliability, generosity, and Good Dad nurturing would likely enable a long-term relationship — even if she couldn't help but check out Mr. Sexy when her estrogen levels were high.

Millions of years later, this trade-off, indeed a strategic female behavior, endures. And we are still feeling the waves of ancient estrous urges in modern times, when during peak fertility women report feeling more drawn to men other than their partners who exhibit characteristics consistent with high-fitness genes — those bad boys in stinky T-shirts.

You might wonder, in the modern world, do these desires translate into women's actual unfaithful behaviors? Rates of female infidelity in Western populations are estimated to be between 20 and 50 percent.[16] Recall, too, that when researchers measure men's symmetry in the lab and quiz them about their sexual histories, men who measure as more symmetrical report having more past partners who were themselves already involved with other men when the affair started — suggesting that women who have affairs tend to go for the good-genes guys. (To be sure, women have affairs for many reasons, including to investigate other long-term partner options.[17] But when at peak fertility, the good-genes scenario might hold sway.)

What about cases of children resulting from affairs? A 2006 study compiled sixty-seven estimates of "nonpaternity" resulting from genetic testing of fathers and their children, either as part of screenings in medical studies or as rates collected from paternity-testing companies.[18] Rates of nonpaternity ranged from .04 percent (in a group of Jewish priests) to 11.8 percent (in Nuevo León, Mexico) — but these rates are only for the participants in the genetics studies, who presumably were high in "paternity confidence," since they freely gave biological samples for testing. Among men who sought paternity testing, who were presumably low in paternity confidence, rates were much higher, from 14.3 percent (in Russia) to 55.6 percent (in the United States). When all of the data were collapsed together, rates were a more reassuring 3.3 percent. However, while the percentage overall might not be very high, we have to remember that these rates occur despite the fact that we have abundant use of contraceptives (and for nearly all samples, it is readily available). That means that rates of nonpaternity could have been much higher for ancestral males. Also consider being one of those one-out-of-thirty men in this situation (or the one in two seeking paternity tests in the United States). *Ouch*. There's a reason why the thought stings so much — the reproductive costs could have made man an evolutionary dead end. Costly indeed, even if it didn't occur with abundant frequency.

Infidelity leading to nonpaternity also occurs in more remote populations. My colleague at UCLA Brooke Scelza studies the Himba, a tribe in Africa, in which social norms allow for fairly lenient views about extramarital relationships. Fifteen percent of women and men report that they know that one of the children living in their household is not the father's biological child. Fascinatingly, and tellingly, all cases of nonpaternity (or nearly all) occur in arranged marriages. When spouses are in "love matches," meaning that they freely chose each other, the nonpaternity rate goes to nil.[19] Even supposed paragons of monogamy, such as "monogamous" songbird females, seem to be fooling around.[20] When biologists peer into their nests and do some genetic testing, they find that about 11 percent of those chicks are being fed worms and bugs by a male who is not their true genetic father.

For ancestral females who evolved to cooperate with male

partners to raise their offspring, unfaithful behavior was perhaps the ultimate trade-off. It might have resulted in better genes but could severely compromise a crucial relationship with their investing mates. Given our evolutionary history, we certainly would expect men to be vigilant about cues suggesting infidelity.[21]

Chocolate Cake for Breakfast

If estrus can trigger such conflict for women, particularly those in relationships — having one thing, desiring another — then what purpose does it still serve? This is a sexual psychology that makes women look fickle — a shift in attraction across the hormonal cycle, from one kind of male to another.

Clearly, estrus served a purpose in animals that still pays off to this day, a way for females across many species to secure good genes for their offspring. But what about humans, particularly in our modern world, with reliable contraception and modern medicine that makes life much less perilous for any offspring we do have? If a long-term mate doesn't offer good gene prospects for offspring, most women won't risk an affair and seek out a symmetrical alpha male purely for his genetic prospects for her children. (Well, some may. . . .)

Perhaps shifts in female desire across the cycle are merely vestigial, the psychological equivalent of our tailbone. We don't need them anymore to produce offspring who survive, but we're saddled with these ancient preferences. Yet, for these estrous features that are a remnant of our prehuman past, they certainly drive a lot of our modern behavior — and not just our sexual desires. As you'll see in later chapters, estrus can impact women in other ways, for example, by blunting our risk-taking behavior (and keeping us safe).

Chocolate cake first thing in the morning sounds like a great idea to me when I wake up starving, but that doesn't mean it's a good breakfast. I know I will crash, and if I ate it on a regular basis the outcome would certainly not be good. Similarly, we don't want to blindly follow our estrous desires — the bad boy may look good, but that doesn't mean he *is* good for you. In both cases, the body is saying, *I crave this!* But if you understand where the craving is coming from — and that it's fleeting — chances are you'll make better choices, whether it means indulging those desires or passing them over as an ancestral relic that is best ignored.

5

Mate Shopping

IF YOU WANT TO survive in this world, don't leave your house. Ever. Surround yourself with creature comforts and a steady supply of food, water, and Wi-Fi, and avoid going out, where risk factors and threats lurk. Sure, you can go in the backyard, but if you want to survive, you are better off indoors and alone, where you can live a long time if you're reasonably healthy and a bookshelf doesn't fall on your head. Say good-bye to things that will put your life in jeopardy, or end it — contagious people, grizzly bears, drone strikes, bad guys with handguns, redback spiders, motor vehicles — they can't kill you if you avoid close contact. (Well, maybe the drone can.)

And sex. Definitely say good-bye to sex and its ability to spread disease. In the animal kingdom, where the only safe sex is no sex, fooling around certainly has a price. Consider the wild Soay sheep that make their home on a remote Scottish island, where winters are brutal and food is scarce. They seem to fall into two categories — very sexually active and somewhat sick, or less driven to reproduce and healthy. The randiest among them produce many offspring before falling prey to the harsh elements or succumbing to a parasitic worm that is particularly deadly to the Soay sheep, which sacrifice immune resistance for fast reproduction. The hardier if more standoffish members of the population live longer but produce far fewer offspring.[1]

Scientists have discovered that those long-lived Soay have high levels of antibodies that protect them from the immune-suppressing parasite that wipes out neighbors one hillside over; healthy sheep pass this robust trait on to the next generation. Even though half the population regularly gets killed off by a nasty worm or the winter weather, the healthy ewes and their fit-gene offspring are able to carry on — so the species endures. The Soay are either building up their immune systems and conserving resources while reproducing with caution, or they're traipsing all over the rocky isle, mating and making lambs like it's the end of the world (and for that half, it is).

The fate of any new lamb is partly determined by parental mate selection — that is to say, either the lambs inherit protective antibodies from fit parents, or they're at high risk for contracting the disease. But the fate of a Soay sheep, particularly a female, is also determined by how she spends her time. Is she investing her resources in mating and raising her young, thereby taking related risks that cause her to grow weaker and fall prey to the parasite? Or is she spending less time in taxing reproductive tasks early in life, thereby conserving energy and preserving her immunity, allowing her to live longer, grow and strengthen her body, and have her offspring later in life?

Like the Soay, you could live fast and die young, or you could hunker down and think twice about answering the door on Saturday night. Go out and be social... or stay safe and go to bed early, alone. Few young and healthy men and women are going to hide themselves away like hermits, particularly during their reproductive years. If ancestral humans had avoided all risk, we would have hit a serious evolutionary dead end.

It is well established that across many species, searching for a mate and producing offspring involve taking a certain amount of risk, particularly for the female (with the infamous exception of the praying mantis; the male literally loses his mind during sex when the female bites his head off after mating). For modern humans, mate shopping may not be quite so life threatening as in other species, but when we put ourselves "out there," we are definitely less safe than if we just stay home. Yet because of the obvious problem with the staying-home survival strategy (see: Evolutionary Dead End), we forge ahead, get up and get dressed, go out, and meet potential mates. In doing so, we're making the same evolutionary trade-off — between survival and reproduction — as the earliest humans did, and as many other species have done for billions of years.

Over and over again, trade-offs emerge as a theme in the evolution of estrous desires. We've just seen a prime example: the trade-offs between choosing Mr. Sexy and choosing Mr. Stable. But that particular trade-off pertains to actual mate selection — as in, What type of male might father the offspring? Before females can choose (and whether or not they act on those desires), they have to put in the effort to gather up potential candidates. And mate shopping, as it turns out, is very time-consuming, packed with key choices, and full of risk (including parasites — and worse).

Fortunately, our hormones guide us, keeping us on task as we seek to make good sexual decisions across the fertility cycle — and helping to keep us safe (at least from certain risks, including undesired mates who might be coercive).

Prime Time: How Will We Use It?

Numerous studies have shown that ovulating women walk more, eat less, go out to socialize more, meet more men, dance more, and flirt more — and I'll describe that research more thoroughly in the pages to come. In theory, ovulating women are seeking out Mr. Sexy, with his fit genes. This uptick in increased physical and social behavior isn't merely an increased "sex drive" (research shows that ovulating women usually do not have sex more often with their long-term mates, though they might have sex more with short-term mates).[2] Furthermore, we've put the "frenzied woman in heat" idea to rest; because they are highly selective, women don't always act on their desires.

Still, there is an undeniable surge of specific activities — behaviors that seem to bring females into more frequent contact with males — as estrogen rises and ovulation approaches. Biologists call this phase "mate-search effort." Because the fertile window is thrown open only a few days a month, that ticking sound may be the biological clock, getting ready to ring. When that alarm sounds (signaling ovulation), the choices include:

Get up and go with the estrous flow: That is, we can get active and be social, increasing our chances for being in situations where we may meet potential mates. This hormonal tug is borne out in the research showing that women are more likely to literally roam their surroundings during peak fertility.

Or...

Hit the snooze button: Ignore the you've-got-to-move-it-move-it estrous hormones and use our precious time for things other than a mate-search effort, an exercise that may be particularly illogical if we're not in the market for a partner. There are plenty of other essential, time-consuming (and not sexy) activities that we can engage in — such as eating, sleeping, working, and childcare.

Either way, when estrogen begins its climb and fertility peaks, we make choices about how we spend our time, and if we choose not to mate, the trade-off is no offspring. But we live in modern times and many of us will take a pass on contributing to the survival of the human race. (Rest assured that someone, somewhere, is doing the deed and taking care of it.)

It is fascinating, however, to see how hormones nudge us toward physical activity and other behaviors that could be interpreted as contributing to the mate-search effort. At the same time, hormones may pull us away from pursuits that could be considered the opposite of mate searching, particularly those that are sedentary and solitary, like checking work e-mails at nine p.m. on a Saturday. Not surprisingly, two competing behaviors that emerge most frequently in studies of both estrous animals and humans are the need to eat versus the urge to roam — a showdown that is sometimes referred to (crassly) as "the need to feed versus the need to breed."

Biscuits or Babies

The pedometer study from the early 1970s, which measured the activity levels of women at different points in their cycle, showed a marked spike in physical activity and movement midcycle, during peak fertility. (See "Heat Seekers, Still Seeking," in Chapter 2. There is also a jump in activity just prior to menstruation, which may have more to do with nesting — but we'll get to that later.) Researchers had already demonstrated that estrous lab rats, farm animals, chimps, and rhesus monkeys were more active — and with the pedometer study, human females were shown to share similar cycle-related behavioral patterns. When estrogen kicks in, women get a kind of cabin fever and feel an urge to get out of the house and roam their surroundings. Unlike lab rats, which can't take the dog for an extra-long walk or go dancing, women have lots of options to work off their estrogen-triggered wanderlust. And since they are not under strict hormonal control, they can also overcome the wanderlust.

My UCLA colleague Dan Fessler analyzed numerous studies on "periovulatory activity" to examine whether caloric intake drops among women and other mammals just as their estrus-driven need for mobility surges upward.[3] Many of the studies were not high quality and were compromised by random noise — including inconsistent research methods — that made patterns difficult to discern (the paper was published in 2003, before most researchers got savvy about using hormone tests to verify women's location in the cycle). Nonetheless, he found that there was a drop in women's calorie consumption just prior to ovulation — when fertility was high. Later, up at my sister campus UC Santa Barbara, Jim Roney and his student Zach Simmons followed up with better measures and repeated hormone tests and found a pattern that was stunningly similar to the one that had been seen in rhesus monkeys.[4] (As Jim says, the rhesus monkey dependent variable — biscuits eaten — should win the prize for best dependent variable in behavioral science. It is so concrete, and what scientist would not want to have "biscuits" on his or her y-axis?) The following graphs, which compare the self-reported eating habits of ovulating women in Roney's study with the number of biscuits eaten by rhesus monkeys in estrus, are almost mirror images of each other:

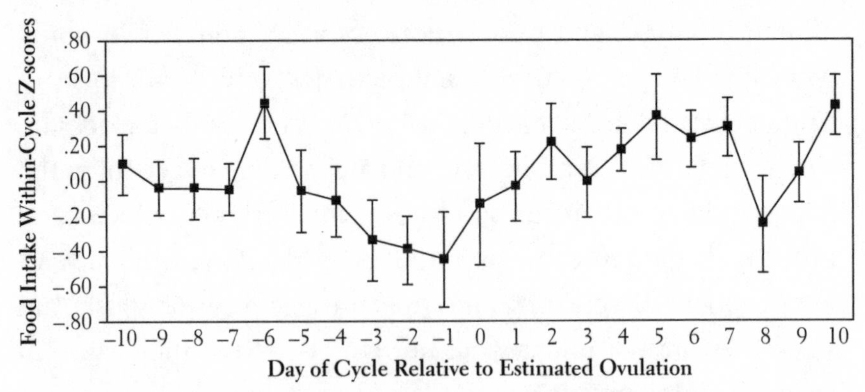

Women's food consumption is shown on the vertical axis (scores above zero mean that food intake is above a woman's average for the month; below zero, below average for the month). The fertile window, approximately Day –4 through Day 0, shows that women's food intake takes a dip when they are most fertile.[5]

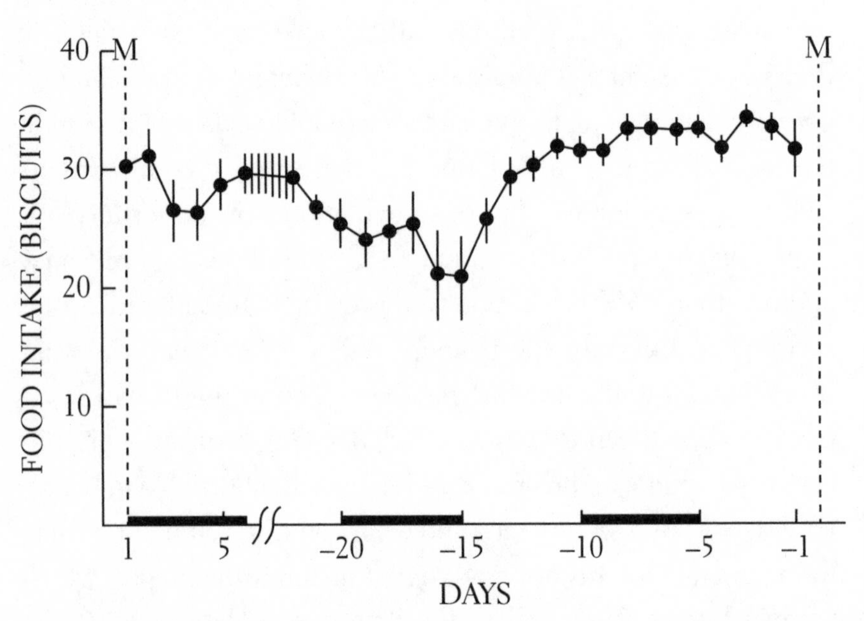

Average number of monkey biscuits eaten across the cycle. The far-left side of the horizontal axis counts forward from menstrual onset (Days 1 through 5). The rest of the horizontal axis presents days before the next menstrual onset, with Day –15 as the approximate day of ovulation.

As Fessler observed in his article, there was good evidence that a cyclical drop in food intake happened. But there were few plausible explanations for *why* it happened. There are no known biological reasons for caloric needs to decline *specifically* during the fertile window. A woman's metabolism is higher after ovulation while the endometrium, which is energetically costly, is preparing for the possible implantation of an embryo.[6] But this fact predicts that the food intake will be lower during the entire follicular phase, starting with the day of menstruation, not just on fertile days.

If anything, we might expect that when the body expends more energy, the desire and need for calories climb. If you're having a burst of physical activity, you'd normally build up an appetite. Apparently, however, this is not the case during the fertile window.

The title Fessler chose for his article nicely sums up his findings: "No Time to Eat." Fessler concludes — and I agree — that "natural selection has taken account of the fact that, during the fertile period, females have better things to do than eat." Namely, engage in the mate-search effort.

Let's Dance: The Human Mate-Search Effort

What does the mate-search effort look like in real life? In 2006, I published the results of research that my colleague Steve Gangestad and I undertook in an attempt to answer that question.

It was work I had begun in graduate school, after brainstorming with my mentor, David Buss. David and I both agreed that it did not seem plausible that women's mating behaviors would remain constant across the cycle (or vary depending only on social circumstances), which was the standard view at the time. It seemed far more likely that women's behaviors would change depending on whether they could become pregnant (when that window is thrown open for those few crucial days). So, we developed many ideas about how human females might shift their efforts and interests depending on their cycles — and how their male partners, in turn, might respond. We set out to do preliminary tests of these ideas, and I followed up in the subsequent two decades. Part of the reason for the project was that I was also really tired of the research I had been conducting on sex differences — there were as many alternative explanations for simple differences between men's and women's behaviors as there were people willing to express an opinion. (*Oh — it's because of the media! Oh — it's because the penis sticks out from the body and is bigger than the clitoris!*) I felt I was getting almost nowhere in advancing theories about human sexual behavior. I thought that by showing subtle, under-the-radar effects of hormones, I might be able to test some evolutionary ideas in ways that couldn't be so easily counterexplained with mundane excuses.

David and I had been looking at the larger question of female desire and male mating tactics, including "mate guarding" by men: when men behave in ways to suggest they are jealous of other males or to signal sexual possessiveness of their female partners — anything from an arm around the shoulders in public to expressing affection in private.

As part of this research we zoomed in on what activities women engage in during peak fertility, specifically whether or not they seek out situations where they might encounter high-fitness genes and the men who possess them. The participants in our study included heterosexual college-age women in pair-bonded (committed) relationships as well as women without steady partners. Based on what we knew about female behavior during estrus — specifically, the desire to be more active and social — we predicted that all women, regardless of their relationship status, would look for situations where they might meet men, though perhaps not consciously.

Using specially developed daily questionnaires, we had women document their behavior for thirty-five days, encompassing at least one full menstrual cycle. (None of the women were taking any form of oral or hormonal contraceptive.) Among the many questions we asked: Have you flirted with or been attracted to men you know? What about flirting with strangers? Or with male friends, acquaintances, or coworkers? Women were asked about their social activity and whether or not they'd be likely to go out with friends to a dance club or large party where they might meet men.

We knew where each woman was within her menstrual cycle: when she was experiencing PMS, if she was having her period, and where she was within the fertile window. Because we had women's responses each day across the entire cycle, we were able to assess participants' answers and look for links between their hormonal shifts and various sexual and social behaviors—whether the women wished to stay home or go out, what they wanted to do, and with whom. The findings confirmed that during peak fertility, women, regardless of relationship status, were more interested in going out to places where they might meet men (other research, described in "Why Even the Good Girls Flirt with Bad Boys," later in this chapter, shows that women actually flirt more when in the lab with an attractive man). We found that women in relationships noticed more attractive men and flirted more with men other than their partners, particularly if their partners were not, by the women's own estimations, highly sexually attractive.

Does this mean a woman won't meet her soul mate if she's experiencing the hormonal dips and swings that come with PMS or menstruation? Of course not. But once, there was a time when specific behavior during peak fertility served a vital human purpose, and that fits together with other pieces of the estrous puzzle. At any given point during her cycle, an ancestral woman had to accomplish a long list of physically taxing tasks in order to ensure not only her survival but also that of her offspring — for instance, she had to secure food and shelter and take care of the children. But when she was most likely to become pregnant, mate-search effort became a priority. It makes sense that during ovulation she would have shifted her tasks from seeking out nourishment to seeking out those high-fitness genes.

In other words, she had no time to eat.

Female Competition: Mirror, Mirror, on the Wall

As a reminder, Robert Trivers's parental investment theory states that females are pickier than males when it comes to mating. Females will be the more selective of the two sexes when it comes to choosing a mate because they must "invest" significantly more effort in reproduction; this makes males the "low investors" — or, at least, the sex that has the *option* of investing less. Women are also far more limited in the number of offspring they might be able to produce — this makes mate choice a particularly high-stakes game for women. A key tenet of the theory is that the "lower-investing sex" will compete for access to the "higher-investing sex." This drive to compete is why males in so many species seem to be armed for battle. Male deer have large, lethal antlers and females do not; male elephant seals are nearly three times the size of their female counterparts. Human males have nearly 50 percent more upper-body strength than women,[7] perhaps not only because they tended to be the hunters in the family, but also because they were competing with other males for possible mates.

An abundance of research suggests that men are generally more competitive than women. But certainly women can be very competitive in many areas — and not just in obvious ones like sports, business, and politics. In fact, a recent study suggests that a woman will be as competitive as a man *when she is competing against herself* — that is, when it comes to competing against her own past performances, rather than competing against others.[8] She will push herself hard to do better next time, whether at the gym or at the office.

Still, if you focus primarily on outward displays of male-versus-male strength, such as greater height or muscle mass, males appear to be more primed for intrasexual competition. (And let's face it — the phrase "pissing contest" was probably not coined by a woman.) So, does this mean that human females do not compete with each other when it comes to the mate-search effort, as males do? Or is there a key context and time that create competitive urges in women? What about when the reproductive stakes are high — when fertility is at its peak and the "best" men and their fit genes might seem to be in short supply?

Let the games begin.

Round 1: Dressing to Impress—But Whom Is She Trying to Impress?

Research from more than a dozen years ago, when scientists were beginning to delve deeper into the question of human estrus, suggested that women do indeed compete more with other women when they are at high fertility within their cycles, but the work was based on tiny samples and measured competition in a quirky way, by asking women to rate other women's faces. In the earliest study on the subject,[9] women at high fertility tended to rate other women as less attractive than did women at low fertility. In other words, high-fertility females tended to be tougher judges of other women — trashing the competition, essentially.

Later research on female competition is more thorough and convincing, and some of the findings build on an outcome from my own lab at UCLA, in a study I published in 2007.[10] We were primarily interested in changes in women's desires across the cycle. We brought women into the lab at high- and low-fertility phases of the cycle, administered hormone tests, and quizzed them about their relationships (with their own partners and with other men). My undergraduate research assistant at the time, Mina Mortezaie, was interested in whether women's clothing styles would change across the cycle, along with their hormonal fluctuations and sexual behaviors. To gather material, we photographed women in the attire they opted to wear to the lab on that day.

We showed the participants' photos (with faces occluded) to a group of men and women — also students — not associated with the study and blind to the purpose of the study. We asked them to pick the photo in which the woman was dressed most attractively (choosing between the woman's high-fertility photo and her low-fertility photo). I was skeptical. As a longtime college professor (and a former student myself), I know undergrads choose their clothing based on whether they have an exam (comfy sweats), never went to bed (last night's clothing), or have an interview or an important meeting with a professor during office hours (nice outfit, straight from the plastic dry-cleaner bag). Fertility, if it had an effect, would surely be swamped by these other factors.

To my shock, we found that the other undergrads, blind to the women's fertility status, chose the high-fertility photos as the ones in which women were "trying to look more attractive" 60 percent of the time, and the low-fertility photos 40 percent of the time. Twenty percent is not a whopping difference, but it is definitely noteworthy as a surprising demonstration that women had fertility, at least in this modest way, on display.

Women in our study, it seemed, were putting themselves out there through what biological scientists call "ornamentation" — in this case, more attractive clothing and careful grooming. But were they also sending a message to female competition? In the wild, male animals have ornamentations not only to show off to females, but also to signal their dominance and superiority to other males. Those massive and deadly deer antlers are but one example of a feature that serves two purposes: to attract the attention of the opposite sex and to do battle. We concluded women might be doing something similar — using outward appearance not just to attract males but to compete with other females, too. This idea needed further study.

About a year later, my colleagues Kristina Durante and Norm Li and I gathered more evidence through a much larger study conducted at the University of Texas. This time the specific goal was to study clothing choices across the hormonal cycle among a broad sampling of women.[11] Would their dress reflect their hormonal status and availability? We asked female UT students aged seventeen to thirty to report to the lab twice: once on a low-fertility day and again on a high-fertility day, when ovulation was approaching or in full swing. As always, we confirmed hormone levels through testing, and none were taking oral contraceptives. Some of the women were in steady relationships, some weren't, and some had never had sexual intercourse.

The variety of the sample is important: If a woman is involved in a romantic relationship and is having regular sex with a steady partner, how she chooses to present herself may be influenced, consciously or not, by that fact. (*My boyfriend is out of town so I'm not washing my hair, and I'm wearing these yoga pants he hates two days in a row!*) If a woman is available and sexually active, she might dress more attractively day in and day out. If she happens to be single and not sexually experienced, she might choose not to dress provocatively at all.

Once again, participants were photographed in the clothing they wore to the lab. This time, however, they were told to imagine that a friend was having a party later that evening, where there would be a good number of "single, attractive" people. They were given templates of female figures and colored pencils, instructed to draw the types of outfits they would wear to the party, and told to indicate clearly where the neckline on their shirt would start and where the shirt would end, and where a skirt or bottoms would begin (how high or low on the waistline) and where they would end. We transferred the drawings onto special paper that allowed us to calculate the number of square millimeters of exposed skin in each drawing, including on arms, neck and shoulders, upper and lower leg, and so on.

On low-fertility days, the outfits women showed up in, as well as those they drew, covered more skin. And on high-fertility days, the clothing they wore and the clothing they depicted for a night out was markedly more sexy and revealing. (Among high-fertility participants, this "revealing" effect was even more pronounced among women who were experiencing the greatest surge of luteinizing hormone, meaning that their chances of conception were greatest.)

Here is an example of what one participant drew. On the left is the low-fertility night-out-on-the-town ensemble (A), and on the right, the high-fertility outfit (B):

Note the longer skirt in the low-fertility picture, and just one shoulder exposed instead of both. Even the shoes cover up more

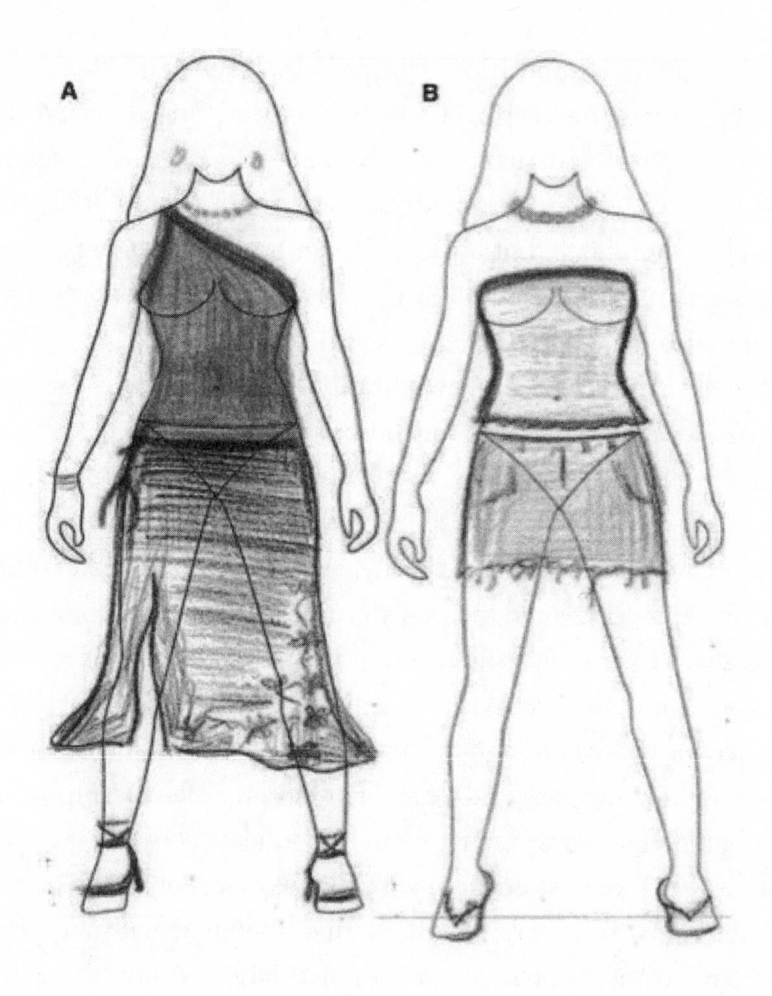

skin. Our findings also showed that clothing choices were associated with factors like relationship status or relationship satisfaction. For instance, women who identified as "sexually unrestricted" wore and depicted more revealing clothing than women who were in committed relationships.

The urge for "ornamentation" was clearly present during high fertility, a sign that women might be trying to signal their status as "on the lookout for mates," a strategic behavior of estrus. As for competing with other women, here's an interesting finding: At peak fertility, sexually unrestricted women — as compared to those in committed relationships, who presumably weren't "looking," and to sexually inexperienced women — wore and depicted the most revealing and sexy wardrobe choices. A flashy necklace paired with a strapless top won't do the same damage as antlers on a twelve-point buck, but if the competition is wearing a twin-set and pearls, and if the conditions on the playing field are just so, then the battle goes to the bared shoulders.

In an effort to detect competition between women, in several studies, my colleagues and I have also asked participants to rate how attractive they felt compared to other women (their "self-perceived desirability"). We asked questions such as: "Compared with most women, how attractive is your body to men?" "Compared to other women, how sexy would men say you are?" We also asked how men might rate their desirability in a variety of relationship scenarios, ranging from casual sex partners to marriage partners.

We found that on high-fertility days, women considered themselves to be more sexy and desirable than other females. In other words, high-fertility women were more confident about their potential success in the mate-search effort than low-fertility women.

In the next chapter, we will see that women actually are more attractive to outside observers at high than at low fertility. So women seem to know, at some level, that the playing field has shifted.

Why Even the Good Girls Flirt with Bad Boys

If women choose more revealing clothing during high-fertility days, do they also choose more provocative ways of communicating with men (i.e., flirting)? A former intern from my lab, Stephanie Cantú, devised a clever, if sneaky, study to investigate whether or not ovulation influenced such behavior.[12]

College women reported to the lab during high and low fertility to interact each time with two men. They had been told that they were participating in a study on how male "identical twins communicate and interact with potential relationship partners." They had also been told that since strangers sometimes get nervous at first encounters, they would live chat via a video interface instead of meeting in person. Imagine Skyping with a blind date, and you get the idea.

But here's the twist. In each encounter, the "twins" were not twins at all, but a single actor playing two different roles: the socially confident and dominant Sexy Cad, as well as the more caring and reliable Good Dad. (Also, the interactions weren't live; the actor had prerecorded scripted Sexy Cad / Good Dad banter, but through a bit of video and tech trickery, the researchers convinced the woman that the man she was talking to was in a nearby room, closely observing her responses through his own video monitor.)

Sexy Cad and Good Dad asked the same kinds of "getting to know you" questions—the "tell me about yourself" and "what do you like to do for fun" opening salvos. But they presented themselves very differently. Sexy Cad came across as dominant, fun loving, and charismatic, if a bit unreliable; the less aggressive Good Dad telegraphed a caring personality and indicated he was interested in family life and a long-term relationship. Each woman had a total of four encounters: two at high fertility and two at low fertility, each time believing that she was video chatting with both twins, one after the other. Afterward, each woman was asked about her interest in the men as potential partners, and later, the videos of her four interactions were shown to a group of raters, who were not told if the woman was at high or low fertility.

At low fertility, women indicated nearly equal levels of interest in each man as a potential short-term partner (i.e., for sex). In other words, a woman was just as likely to go with Sexy Cad as with Good Dad. But at high fertility, the desire for the Sexy Cad shot up significantly, while sexual interest in Good Dad was much lower. Sexy Cad, after all, was showcasing his good genes through his dominant behavior, seemingly leaving his twin brother in the dust. (Even though the "twins" presumably shared all of their genes, the women weren't thinking about that—only that one guy was displaying the more sexy qualities, whereas the other was not.)

Furthermore, the independent video raters repeatedly noted far more flirtatious behavior directed at Sexy Cad when women were at high fertility. (During low fertility, there was not a significant difference in how much women flirted with either man.)

As we've seen repeatedly, estrus seems to provide a hormonal strategy that enables women to be highly selective.

Round 2: Mean Girls or Wise Women?

When women assert themselves, too often the retro and wrongheaded trope is that they're self-servingly sharp elbowed or, worse, bitchy. But they are just as likely to be faulted for being "too nice," code for saying that others will walk all over them when push comes to shove. Still, though women may hesitate to compete with the opposite sex, as discussed they *will* compete with one another, especially within the mate-search effort framework, whether or not a mate is the goal.

A group of researchers wondered about the role that female hormonal shifts would play in same-sex competition, including a woman's willingness to "dehumanize" a competitor.[13] If women were willing to be more intrasexually competitive during high fertility (on top of feeling more attractive), perhaps during this phase they were also somehow altering their views of other women to dehumanize them in some way, thereby recasting them as "the enemy." It's hard, after all, to compete with someone you like, and someone who is like you. On the other hand, it is much easier to wage war on your competitor if you can view her as "other" — even easier if you can view her as less than human.

In the study, researchers asked women at high fertility to assign descriptive words to three groups of people: men, the elderly, and other women. The women were instructed to choose eight terms for each group from the following list. (The researchers purposely chose descriptions that were "animal-related" or "human-related.") Here are the twenty words they provided to the women in the study.

wife	*pet*
maiden	*mongrel*
woman	*pedigree*
person	*breed*
husband	*wildlife*
humanity	*critter*
people	*cub*
civilian	*creature*
man	*feral*
citizen	*wild*

When researchers reviewed the results, they found that participants used significantly more animal-related words to describe other women, while choosing human words to describe men and elderly people. These findings suggest that when a woman is at high fertility, and therefore at her most competitive, she is able to dehumanize her opponent and think of her as a nonhuman "creature," "critter," or "mongrel." As brutal as that sounds, it might help her become even more competitive. Would you rather defeat "humanity" or vanquish something considered "feral"?

It's impossible to know what the women in the study were thinking when they considered the three groups. "Men" could have been husbands, boyfriends, sons, brothers, or fathers; "the elderly" could have been a frail grandma; and "other women" could have been friends, sisters, daughters, mothers, a partner's crazy ex-girlfriend, or a backstabbing female colleague. But there are reasons why women reserved the most dehumanizing terms for other women, and it couldn't be simply due to some monthly nice-girl-turned-mean-girl transformation. High fertility encourages women to behave more competitively. And it has to do with the girl-eat-girl business of mating when the stakes are highest.

Round 3: Why Women Won't Share—Stingy or Strategic?

As we've seen, women at high fertility may be competing with one another for men who satisfy their estrous desires. It appears, though, that they also compete for fundamental resources, not just mates.

Same-sex competition for resources has been observed when women play "the ultimatum game,"[14] a common research technique used to study human cooperation. In the UG, as scientists call it, one person is the "Proposer" (the one with the resources to share, usually money) and the other is the "Responder" (the one to whom the "endowment" is offered — and who says, "I'll take it!" or "No, thanks"). If the Responder refuses what is being offered, then both parties "lose" — meaning that neither the Proposer nor the Responder gets any of the endowment. Let's say the full amount of the endowment is ten dollars. The Proposer might offer the Responder half the money, so they both get five dollars (seemingly a fair deal). Now, let's say the Proposer offers only one dollar (more selfish, but perhaps worth a try in an attempt to get more money for oneself, or simply to deprive a competitor in the game). The Responder could reject the low-ball amount, and neither party gets any of the money. If you are the Proposer, then, the trick is to offer a low enough amount that you get a good economic payoff — but not so low that the Responder rejects your offer, meaning that you both wind up empty-handed.

Would women allocate resources to other women differently, depending on their hormonal phases? My colleagues at UC Santa Barbara Jim Roney and his student Adar Eisenbruch set out to answer that question.

In an earlier study[15] using the ultimatum game, researchers concluded that fertile women were sufficiently interested in depriving other women of financial gain by offering them low amounts of money — an effect that was particularly apparent when the high-fertility Proposers encountered attractive women. One interpretation of this is that Proposers viewed the attractive Responders as rivals for potential mates. (*You're too attractive, which I find threatening at this time, so I'm not sharing — even if both of us lose and I run the risk of pissing you off.*)

Eisenbruch and Roney set out to build on this initial research with some variations on the ultimatum game. Female participants were presented with photographs of other women and were instructed to come up with an offer they would make, from an endowment of ten dollars, to the women in the photos. In addition, each participant made her own financial demand: how much money she'd want from the woman in the photo, again, out of an endowment of ten dollars. (Think of it going something like this: *You look to me like you'd take three dollars, so that's what I'll give you. As for what I want, I'd like you to give me seven dollars.*) In classic ultimatum game fashion, the Proposer still ran the risk of walking away with nothing.

Overall, the findings echoed the earlier research, as well as the other trends in intrasexual competition: At high fertility, women offer less and demand more. Furthermore, women seem to be targeting their competition — attractive women. In the earlier research, low-fertility women were *more* cooperative with attractive women (as well as with attractive males), not less. But at high fertility, when they are most likely to become pregnant, women are willing to withhold resources from potential rivals, even if it means they get nothing themselves.

It's a leap to say that part of the evolution of estrus is women competing for monetary resources at high fertility. My view is that women just feel more competitive during this phase; the most desired men, after all, are a scarce resource. But this spills over into the domain of financial competition. Make of it what you will — high-fertility women compete more, but perhaps only because they are a bit more like their male counterparts at this point in their cycle.

Risk Management: When Women Think Twice

In the same way that we think of men as being the more competitive sex, we also consider them to be bigger risk takers compared to women — and they are. Men gamble at higher rates than women, and not just with money. They drive faster, are more likely to drink and drive (or drink and text), and are more likely to be involved in fatal car accidents than women. (Car insurance premiums for teen drivers are generally much higher for boys than they are for girls.[16])

Women are known to be more cautious (and, statistically, more law-abiding). But there are certain risks they deem worthwhile — including risks they take during the fertile window that they wouldn't take at other times. The mate-search effort nudges a woman to leave the protection of home and venture into nighttime social situations like clubs and parties, when she is likely to choose more revealing clothing and flirt with male strangers, including those whose motives may not benefit her. Women take risks when they grow more competitive with other women. As we've seen, in the ultimatum game, a high-fertility woman is willing to risk the loss of resources and possibly anger her opponent. And, of course, the culmination of the mate-search effort — that is, sex itself — carries its own risks.

Yet, women at high fertility also seem to be highly strategic about risk. Despite the urge to be out and about, there is evidence that the same hormones responsible for flirty behavior and shorter skirts may also cause a heightened awareness that has a protective effect.

The Dark Alley

The mate-search effort carries risks involving sexual predation, but women in estrus seem primed to avoid it. Diana Fleischman of the University of Portsmouth (UK) led a group of international researchers in a study that looked closely at the seemingly contradictory behaviors of ovulating women, and at whether hormonal fluctuations played a role in a woman's avoidance of sexual assault.[17]

Previous research had shown that ovulating women engaged in fewer behaviors that might raise the chance of sexual assault, such as walking down dark alleys straight out of central casting, or avoiding men whom they perceive as dangerous. Yet, women at high fertility also take risks by seeking to be noticed. But are these two sets of estrous behaviors — being risk averse and taking some risks — truly contradictory? Or do they fit together with the idea that women are highly selective when it comes to choosing mates?

Fleischman and her team analyzed the responses of women who were asked about a wide variety of "risky" behaviors. She pointed out that in the previous studies researchers who asked women to rate risk tended to lump a range of classic "risky behaviors" into one big category. For instance, pulling into a remote rest stop on the highway and going into an unfamiliar dance club in a seedy neighborhood could both be considered "risky" actions for a woman, but one is not related to mate search (she is falling asleep at the wheel and needs to get off the road) and the other is (she wants to go out dancing and meet men). Fleischman's study sought to determine more precisely whether women at high fertility would avoid certain types of risk related to potential sexual assault, not just risk in general.

Researchers presented participants, college-age women at

high fertility, with a "risky activity inventory" questionnaire, which listed a wide range of activities they could do on their own, with a friend, or with a male they were interested in as a short-term mate (a date) or a long-term mate (a steady romantic relationship). An ordinary activity would be presented in a variety of scenarios, and respondents would rate its risk level in each instance: for example, going outside during daylight hours to empty the garbage (1) alone, (2) with a friend — female, (3) with a friend — male, (4) with a date, (5) with a regular romantic partner. Then, what about doing this same activity after dark? Would the perceived risk level rise? Not surprisingly, it did. Researchers, therefore, weren't just getting one answer on whether something was risky; they were getting a more comprehensive result.

In addition, because a goal of the study was to see how the menstrual cycle might impact the risk of sexual assault, researchers asked the same women why they felt certain scenarios were riskier or more fear provoking than others. For example, going to one's car in a dark parking lot or riding the bus alone at night caused fear, but why? When participants were at peak fertility, they mentioned fear of rape or sexual assault — as opposed to theft or being harassed in a nonsexual way — more than women at low fertility did.

Women assigned the highest level of risk to situations where they were "alone or interacting with strangers when accompanied by friends or kin." An ancestral woman who was young and fertile probably stood to lose more than an older woman if she were the victim of rape and unwanted pregnancy. If she were to be impregnated by a man she didn't choose, chances are he would not have had the high-fitness genes she might be in search of, and having a young child could foreclose other valuable mating opportunities.

Fleischman's study, like the research before it, shows that when women are at their most fertile and the chances of conception are high, they are more cautious when it comes to non-mating-related behaviors — including walking down that dark alley.

The Dangerous Stranger

While high-fertility women are seeking male attention, they seem inclined to avoid activities that expose them to the wrong kind of attention. But what happens if they can't avoid the risk and encounter an actual threat, in the form of a sexual predator?

One much-discussed study had women read stories about the threat of sexual assault while holding handheld dynamometers (devices that measure handgrip strength).[18] The stories involved scenarios like a woman walking alone to her car at night and sensing that someone is watching her. The women who were in the fertile phase, the study claimed, had stronger grips, the implication being that fertile females who felt threatened had an uptick in physical power and could fight off sexual coercion (there was also a small rise in testosterone during the fertile period). That said, a well-timed knee to the groin (administered strategically, at any time of the month) is probably a more valuable self-defense skill than a really firm handshake.

When women are feeling vulnerable and come face-to-face with a threatening male, there is some evidence that suggests a shift in their visual perceptions of certain men, and those shifts seem to be more apparent at high fertility.[19]

In a study of more than six hundred women, researchers showed women "mug shots" of two men, one of whom they described as having been convicted of tax evasion and the other as having been convicted of aggravated assault. (The men weren't actually criminals nor were the mug shots real; they were composite photos of other men.) They were also shown silhouettes (faces obscured) of male figures with various body types, ranging from short and slight to taller and more muscular. Using that scale, women were asked to estimate the physical size of each of the two "criminals."

The women had also completed questionnaires rating their levels of fear regarding a variety of crimes, such as mugging, car theft, and sexual assault. In addition, researchers collected data on where the women were in their menstrual cycles.

The results showed that when chances of conception were highest, women who were most fearful of sexual assault estimated the body size of the violent offender to be taller, larger, and more muscular than that of the tax evader. This might seem to contradict women's estrous preferences for the Sexy Cad, but it's all about context here. When a woman is clued in to danger of possible sexual assault (and her ability to freely choose her partners), the big guys are not going to be the most appealing. These findings fit with the idea that women, in general, are cautious about risk and possess a kind of internal alarm system. And at high fertility, that system is fully armed.

The Ick Factor (or, "Can I Talk to Mom?")

Estrus can help a woman steer clear of certain types of risk and — even if offspring are not a goal — keep her eye on the mate-search-effort prize: good genes. However, there is one source of very (very) bad genes that women frequently encounter during the fertile phase, though on the surface these males are not menacing: their fathers. There is no easy way to talk about inbreeding risk among humans (trust me, I've done this in lectures for years and it always makes everyone squirm). But it's a very important part of the estrous discussion. Will women avoid this particular risk — from male kin — during peak fertility, in the same way they avoid dark alleys and dangerous strangers?

I've conducted and written up many studies over the years, but one of my favorites — despite the ick factor — is what I've come to call "the cell-phone study."[20] Together with Debra Lieberman, my dear friend, and Elizabeth G. Pillsworth, my first graduate student, I wanted to learn whether or not "inbreeding avoidance" was a hallmark of human estrus. Scientists have observed this avoidance in animal studies involving cats, horses, voles, and mice, among others.

Females will avoid male kin when they are in heat, which makes tremendous sense from an evolutionary point of view; offspring from related mates are often unhealthy and have higher mortality rates. Harmful but rare recessive genes can hide in the genome because their effects are rarely expressed. It typically takes two of them, passed on to an offspring, to see their effects. Because genes are shared within families, harmful recessive genes will be too. As a result, across species, there has been strong selection over evolutionary time to detect kin and avoid mating with them. (And this is why the thought of it is so disgusting. *Ew! Ick! French-kiss my brother? I can't believe you weird psychologists would even ask me that!*[21]). In nonrelated mates, a "good" dominant gene from one parent is likely to override the "bad" recessive gene from the other, resulting in healthier offspring. There may also be some benefits related to immunity: Nonrelated parents could each possess different immune-related genes and so could pass along a broader range of immunity to their offspring.

But the idea of testing inbreeding avoidance among humans? How would that be possible? We know from other studies that, at peak fertility, women's feelings of disgust increase at the thought of incest or bestiality (the disgust is a given, but during the fertile phase it spikes higher). Still, those studies did not investigate whether or not women would make efforts to avoid male relatives during estrus. And so, the cell-phone study was born.

Cell-phone records span one month, the rough equivalent of one menstrual cycle. We thought: How convenient! Our study participants — female UCLA students — provided cell-phone bills and detailed information on their cycles. The billing information was carefully itemized so that we could identify whether a participant was separately calling her mother or her father.

At high levels of fertility, women called their dads less often, and if their dads did call them, they hung up the phone more quickly. We saw the opposite pattern for calls to and from moms — more calls and more time spent on the phone with moms when daughters were at high fertility. The study confirms what has been

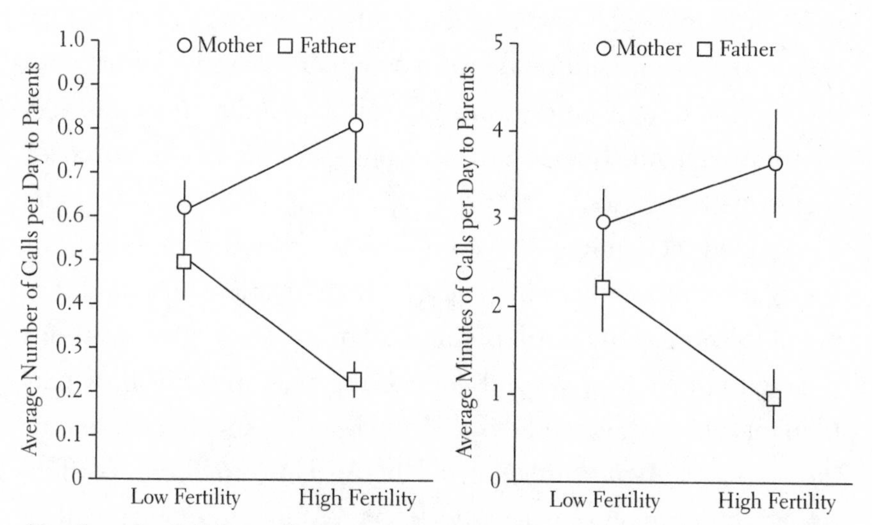

Number of cell-phone calls from women to mothers and fathers (left panel) and duration of calls received from mothers and fathers (right panel). The pattern shows that on high-fertility days of the cycle, women call their dads less often (and moms more often) and end calls with dads sooner when their dads call them (as compared with calls received from moms).

observed in nonhuman females: During estrus, women avoid males who would make for very bad mating decisions, indeed.

Our study was published in one of the best empirical journals in the field of psychology — *Psychological Science* — but it's not as frequently cited as my other work, perhaps because the hypothesis that women are avoiding inbreeding genuinely makes people uncomfortable. It's hard to believe that inbreeding with close family members could ever happen, and given that women are strongly motivated to avoid it in general, there's some suggestion in our data that men are more blasé about sex with relatives and in some circumstances could even seek it out. (I was excited when the study was published. I thought the methods were clever and the results looked so clean. But I certainly didn't call all of my relatives to gush about it. *Awkward!*)

As some readers might be relieved to learn, there is a somewhat more palatable explanation for the women's behavior that points to a more traditional father-daughter dynamic: Perhaps on their fertile days, women want to avoid paternal control of their mating behavior. (Papa, don't preach — it's all in the name of securing good genes.)

Women generally prefer to talk to their mothers rather than their fathers about certain topics, such as relationships, and it is possible those mate-search-related conversations crop up more during high fertility. We did collect information from study participants about the quality of their relationships with each parent too. Only women who reported being close to their mothers upped the number of calls as well as the duration of each call to mom during high fertility. Women who reported being close to their fathers, however, did no such thing: In fact, talk time with dad was reduced.

Even if our cell-phone study sets off a collective "yikes," it's simply what happens when smart women use their smartphones — and their hormonal intelligence.

The research on female competition and risk management is somewhat spotty — a study here, a study there, and often with small samples and no hormone tests, suggesting pretty noisy data. Although the theory guiding the work fits nicely within the larger framework of women's estrus, I have had my share of skepticism about just how strong our conclusions should be as scientists. However, a PhD student from France, Jordane Boudesseul, who spent time working in my lab, compiled the results on risk taking, including unpublished results, in a meta-analysis.[22] He found that, when taken together, the findings held up, even when he used various means to correct for factors that could paint a misleading picture, such as a bias in the studies that are published (which appear to show something interesting) versus those that are not (which typically seem to show nothing). I found this reassuring, but I still think we need more research in these areas, particularly given the implications for understanding important aspects of women's lives, such as how they compete for resources and how they avoid sexual risks.

Mate-search effort is capable of sending women on a journey where decisions involving survival versus sex are compressed into a very narrow window of time — just a few days a month. Women in estrus make choices about how they'll spend their time, where they'll go, and whom they'll seek out — choices that look different at other times of the month. They may engage in behaviors we tend to associate with masculinity, such as competition and risk taking, yet they are also more cautious — particularly if they feel vulnerable — and more attuned to certain risks that may threaten their access to good genes.

These mate-search behaviors during high fertility can be observable, like flirting and style of dress, and might provide clues to other people about a woman's fertility within her cycle. These clues are not perfect indicators of impending ovulation, given the many factors beyond a woman's hormonal state that affect mating motivations, but they are there. There are other potentially detectable changes in women that other people might be able to detect too, though they are not as clearly tied to women's mating motivations, and probably operate mostly outside of a woman's conscious control — things like changes in a woman's scent, voice, and face. As you are about to find out, these changes could play a role in women's interactions with current and prospective mates, perhaps just as much as putting on a short skirt and going dancing.

6

The (Not Quite) Undercover Ovulator

THE NEXT TIME YOU go to the zoo, do some people watching — particularly when you get to the chimps and orangutans, our closest primate cousins. Note the reaction that humans have to the animals, especially when the females are exhibiting their hormonal status: Primates will have brightly colored genital swellings that (roughly) coincide with high fertility, a quality that males find very attractive. Some humans — especially kids — will have predictable reactions to this particular "display."

Mommy, what are those red things on her bottom?

Oh, she's just happy! Come on, let's go look at the baby giraffe!

When I first started researching human estrus, the belief was that women did not exhibit any outward signs of fertility, or "ovulation cues." Most animals, by contrast, did not seem at all discreet about their readiness for mating and conception. I show slides when I give the ovulation-cues lecture to my undergrad students — pink and red swellings and all — and, like some people at the zoo, they probably would rather move on to other exhibits. They are relieved when we discuss the fact that humans, of course, do not possess this feature. But the reaction to these alien sexual swellings proves an important point: *Beauty is in the eye of the beholder*, or as one of the most prominent evolutionary thinkers of our time, anthropologist Don Symons, said, "Beauty is in the *adaptations* of the beholder." So, what serves as a sexually attractive ovulation cue in one species will not necessarily be the same in another, and just because humans lack sexual swellings does not mean we lack cues of ovulation.

Over the last few decades, we've discovered that women do engage in behaviors that appear to be related to their hormonal phases, such as choosing sexy and revealing clothing and taking extra care in their grooming at high fertility. Women are even known to wear the colors red and pink as ovulation approaches,[1] like our animal cousins who regularly fly the same colors (granted, in a more flamboyant way).

Such ornamentation is a hallmark of the mate-shopping behaviors we just explored, along with women being more social, flirtatious, intrasexually competitive, and selectively risk taking, as we noted in Chapter 5. But there is another set of possible ovulation cues that operate outside of women's conscious awareness (and probably mostly outside of women's control). These cues are subtle — probably far more so than those of our animal kin. For reasons we're still discovering, women evolved to keep fertility status *almost* under wraps, but the ovulation cues are there, as we will see, if you know where to look.

Scent Cues: Strong Enough for a Man, but Made by a Woman

Do you know the old commercial for Secret deodorant? It was *strong enough for a man... but made for a woman.* That slogan is an unwitting twist on hormonal intelligence. Ovulation cues are transmitted through female body odor — scent subtly made by a woman, but just strong enough for a man to detect.

During the average twenty-eight-day cycle, hormonal fluctuations cause women's vaginal odors to change; at high fertility, the scent is apparently more appealing to males.[2] Let's not forget those pioneering 1970s researchers — see "Heat Seekers, Still Seeking" in Chapter 2 — and the women who contributed "samples" for sniffing in one of the first studies to confirm this. (However, let's please forget whoever created the "feminine" deodorant spray, FDS — aka, Fresh Down South. It debuted in the late 1960s, but fortunately the first of many editions of *Our Bodies, Ourselves* was imminent, helping to counterbalance, though not eliminate, the popular obsession with these "hygienic" concerns.)

Males prefer the scents of high-fertility females to such an extent that some male animals — for instance, cats, dogs, hamsters, monkeys, and cattle — will attempt to mate with dummies that have been rubbed with the scents of estrous females. Even though some people still resist acknowledging our commonalities with the animal kingdom, when it comes to mating behaviors, the evidence is clear that human males also prefer the high-fertility scent. We know this not just from the seventies tampon study but from other studies, too.

But many of those early human-scent studies were small in scope and did not track women's fertility with accurate hormonal testing — that is, carefully testing urine, blood, or saliva for the presence and levels of specific hormones in the body, such as luteinizing hormone or estradiol (estrogen). Nowadays, anyone can walk into a drugstore and purchase an over-the-counter ovulation kit to track the surge of luteinizing hormone and pinpoint peak fertility. But decades ago, testing methods were less accessible and had yet to be refined.

One common method for determining a woman's fertility was — and in some cases, still is — for a researcher to simply ask the date of her last menstrual period and to estimate the day of ovulation by counting backward and assuming a twenty-eight-day cycle. But unless a woman keeps meticulous records of her menstrual cycle and has an exact four-week cycle with Day 14 as the Holy Grail, then relying on memory and counting days can make for murky data.

Instead of working with self-reported information and estimation, it's better to gather accurate data by testing for the presence and levels of specific hormones. My students and I wanted to investigate the importance of scent as an ovulation cue in human mating behavior. But we wanted to do so with more rigorous methods.

And that explains why we found ourselves collecting and bottling women's armpit odors.[3]

Stinky T-Shirts: They're Not Just for Men

As the science has shown, a woman's regular romantic partner might find her vaginal odors more appealing at certain times in her cycle — though unless he's paying close attention to the calendar (and crushing on Jane Goodall reels), it's unlikely he'll consciously recognize that his increased attraction is in sync with the approach of ovulation. But hormones travel throughout the entire body. If they affect vaginal odor, they could certainly affect more general body scents. And those might be even easier to detect on a day-to-day basis, particularly for male partners who sleep in the same bed as a woman.

But how about in a lab setting? Would a woman emit subtle ovulation cues through body scent that even strangers might detect? As others had done in previous studies, we collected female scent samples during both high- and low-fertility days and asked men — who did not know the women — to rate them. But we did it with greater rigor (and, at times, with a sense of humor).

Our "scent donors" were female university students who were not on contraceptives (which would interfere with normal hormone levels) and who were nonsmokers. Since the odor of cigarette smoke has a habit of clinging to everything and everyone, we eliminated smokers because they might submit contaminated samples. In addition to interviewing women about their cycle length and regularity, we confirmed high- and low-fertility status by measuring the level of luteinizing hormone in urine samples. Luteinizing hormone (LH) is the hormone I called the Bungee Jumper in our earlier discussion of the cycle, because it spikes quite high twenty-four to forty-eight hours before ovulation, before dropping steeply as the fertile window closes. Urine testing for LH is estimated to be about 97 percent accurate, so it is a far more reliable measure of ovulation than counting days.

To collect scent samples at specific high- and low-fertility points, we had women wear gauze pads under each arm for twenty-four hours before reporting to the lab to submit their samples. We didn't just hand them pads and say, "Here, stick these under your pits!" Instead, about three days before their predicted high- and low-fertility dates, they came into the lab and we carefully conveyed instructions for taping the pad in the armpit area in a specific way so that it would come into contact with the skin. In addition, we had a strict "wash-out" phase they had to abide by: washing sheets and clothing they would be wearing with fragrance-free laundry detergent; no use of scented shampoos, skin lotions, or bath soaps; obviously no deodorants, antiperspirants, or perfumes. (This will sound similar to the "stinky T-shirt" protocol, described in Chapter 2, of Steve Gangestad and Randy Thornhill; I have learned that UCLA students shower all too frequently with fruity, flowery bath products.)

For that twenty-four-hour period we also essentially imposed a policy of no sex, drugs, or rock 'n' roll: no sexual activities with another person; no sharing of a bed with stinky boyfriends or pets who could contaminate the samples; no use of tobacco products or recreational drugs; no booze; no hanging around in spaces with strong odors (avoid the apartment of the friend who lights up, whether it's incense, cigarettes, or a bong, and stay away from smoke-filled bars and parties). Finally, we asked them to avoid consuming pungent foods such as garlic and pepperoni. We took away their pizza.

After twenty-four hours of not living like normal college students, the women removed the pads and placed them in sealed plastic bags before delivering them to the lab. We froze the samples upon receipt for about three weeks at –17 degrees Celsius to preserve the odors, until it was time to rate them. We also debriefed our participants, to confirm they'd followed instructions and to determine if they'd engaged in any activities, such as a strenuous workout, that might affect their scent. Eventually, it was time to pull the samples out of the deep freeze and call in the sniffers.

Our panel of male raters consisted primarily of other university students. Once again, we eliminated smokers from the study. A smoker is twice as likely to have an "olfactory deficit" (a crummy sense of smell) as a nonsmoker, and we wanted guys with good noses. To set up the rating sessions, we brought the samples to room temperature in small plastic bottles, putting each high- and low-fertility sample from one woman into its own bottle. (There was also a third bottle, which contained a sample from the same woman, randomly assigned to be at high or low fertility.)

Each man was presented with three bottles and told to take a "hearty sniff" of each, then rate the scents. He did not know that the three samples in his set of bottles came from the same woman. We asked participants to rate the scents in terms of pleasantness, sexiness, and intensity. We also asked them to guess at the physical attractiveness of the woman, based on her scent, on a scale of 1–10, with 1 being "very un-sexy / pleasant / intense / physically attractive" and 10 being "very sexy / pleasant / intense / physically attractive."

And . . . the highest marks went to the high-fertility samples. More often, men deemed the high-fertility scent more pleasant and sexier (and less intense) than the low-fertility scent, and they guessed that the high-fertility donors were very physically attractive. They ranked the low-fertility scents as less pleasing.

These results confirmed what we knew from animal studies as well as from the limited amount of research on human ovulation cues: Men can detect scent cues that coincide with ovulation, and they find them more appealing. But we showed that it isn't just sexual partners who are capable of detecting these cues. Other men (and perhaps other women) can pick up on them as well, at least from up close and when body odors are uncontaminated by other scents.

For their part, women are not limited to their regular sexual partners in terms of who else may find them attractive. I think it is most plausible that regular partners are the primary others who can detect ovulatory shifts in odors, but I would not eliminate the possibility that through scent cues, a woman may attract other potential partners.

When Men Finally Get a Cue: The Male Response

Human ovulation cues such as a change in a woman's scent are not obvious, and only some men will be able to detect them. (There are potential evolutionary reasons why these scent cues are subtle, meaning that not *all* males will get the memo, and we'll get to that later.) To be clear, a fertile woman isn't sending out the equivalent of a hormonal group text every four weeks: *What up? I'm ovulating!* Even the male who successfully detects a cue won't decode its meaning (unless he's using a fertility tracker app, but that's between him and his partner). His brain isn't thinking, *She's fertile! Offspring! Ding, ding, ding!* or *Danger! Danger! Condom! It's more like, Mmmm, I* like *her.*

My research shows that women who are at peak fertility feel more physically attractive at this time, and those with romantic partners report that their boyfriends or spouses are more likely to mate guard during the fertile window (becoming more jealous and possessive).[4] But those findings were based just on what women were reporting about how their partners behaved. Perhaps her belief that her sexual appeal was triggering jealous and possessive behavior in her partner was a one-sided affair. Or perhaps she interpreted his behavior differently at high fertility because she was noticing other possible mates, and the arm draped around her shoulders felt like a possessive maneuver rather than a show of affection. We wanted to measure men's behavior toward their partners in an objective fashion in the lab to see if mate guarding when their partners were at high fertility would be consistent with what women had reported in earlier studies.

Slow Dancing for Science

As a scientist with a lab at a large university, I can confirm that it is generally much easier to recruit female college students to participate in research than to find willing males. (As a side note, some researchers have observed that at high fertility women are more likely to volunteer for studies — perhaps they're feeling restless and want to put that energy somewhere, though my lab has never asked women to run on a treadmill!) For this study we needed couples. We knew men were more attracted to scents collected from women at high fertility, but did the hormonal cues those scents contained trigger them to engage in mate guarding? Our hope was that women would give their partners a little nudge to join the study, and fortunately they did.

The "challenge hypothesis" is well-known in animal studies, where researchers examine male behavior toward their rivals specifically when their mates are at peak fertility. In male animals — ranging from birds to primates — testosterone levels have been shown to increase when (1) females are at high fertility, and things heat up (sex), and (2) male rivals come calling, and things heat up (fights). We wondered if the challenge hypothesis would be supported among humans, and we came up with a way to test it — minus the sex and violence, of course. Our study[5] would be the first direct test of the hypothesis in men and women.

We brought each couple into the lab two separate times, at high- and low-fertility points for the woman, and created a range of activities for the partners to do together and separately at each session. First, though, we confirmed the woman's fertility by measuring the level of luteinizing hormone, and we measured the man's level of testosterone through saliva testing (a standard and reliable measure), at three time points in the study: upon arrival at the lab, before the competition manipulation, and fifteen minutes after.

In each session, after the initial saliva sample was collected to establish a testosterone baseline, the session began with a ten-second hug, followed by "couple interaction" tasks.

In the first task the couple picked a song from a playlist we provided and slow danced. Slow dancing, as any eighth-grade boy will confirm, is a great way to get in there and smell a girl's neck or otherwise check her out, and that's precisely what we wanted the man to do — to expose him to potential cues of his partner's fertility within the cycle.[6] After they finished, we had them take "cute, couple-y photos," photo-booth style (with the hopes that he'd keep getting exposed to cues of her fertility).

Remember that these were committed romantic partners, so the activities they were engaging in were fairly normal for any couple — even if they were unfolding in a research lab and not at home. We gave the couple privacy to complete the couple-interaction tasks by putting them in one of the research cubicles in my lab and closing the door — these are small spaces that barely accommodate a desk and a chair, but that was precisely the point!

After their lab date, we separated the couple and took a second saliva sample from the man. Next, we showed the man photographs and biographical information of ten other men, each of whom fell into one of two distinct categories: "high competition" hunks or "low competition" wimps.

We were now testing the "challenge" part of the challenge hypothesis; we wanted our male participant to perceive these other men as rivals. To that end, we informed him that his girlfriend was also now reviewing the same pictures and rating the attractiveness of the same guys, who we said were also students at UCLA. Just to give you a sample of our high- and low-competition rival males: a square-jawed fellow (who was undoubtedly symmetrical) described himself as a leader of others, someone who was always being asked "to run for this office or that." His round-faced bro, on the other hand, said in his bio that he preferred to be "behind the scenes" and had a nice-guys-finish-last quality.

The men rated the males in the photos for competitiveness,

dominance, and physical attractiveness. Not surprisingly, participants scored the hunks as serious competition and the wimps as low threat. After the men completed this final task, we took a third saliva sample.

So, to re-create a test of the challenge hypothesis with humans, we collected (1) baseline levels of testosterone; (2) testosterone levels after the men were exposed to an ovulation cue through close couples interaction; and (3) testosterone levels after the men viewed potential rival males. We brought couples into our lab twice — at both high and low fertility in the women.

We predicted an increase from baseline levels of men's testosterone during the high-fertility session for men exposed to the highly competitive male rivals — and that is what we saw. Men viewing the high-competition males showed a testosterone response to these possible rivals that was greater when their female partners were at high fertility versus low fertility. Men exposed to average male rivals showed no such difference across high- and low-fertility sessions.

It's clear when you look at our findings that it's hard to ignore our animal natures. And it's especially hard to ignore the facts of evolution. Though men cannot pick out the ovulating woman from the crowd just by looking, males in nonhuman species evolved to detect and act upon ovulation cues, both to produce offspring and to ward off competition. While animals are known to act upon hard-to-miss cues like sexual swellings, subtle ovulation cues such as scent can also clearly play a major role in driving the message home.

As our study showed, ovulation cues, once thought to be completely concealed in humans, are regularly emitted by women as a result of hormonal fluctuations and are detected by men. As it turns out, other women can detect them as well.

Men Are "Hormonal" Too

We don't think of men as having hormonal phases the way women do, but in fact they do have daily testosterone fluctuations. For men, testosterone levels are highest early in the morning (in case you're wondering, *that's* why you're getting poked in the back before the alarm goes off). They decline nearly immediately thereafter—by 60 percent or more within thirty minutes of waking.[7] Testosterone plays a major role in male physicality—such as maintaining muscle and sex drive. It makes sense that ovulation cues might impact the level of testosterone.

To test this, in one Mexican study men rated the appeal of underarm and vulvar scents of women at both low and high fertility and answered questions about their interest in sex.[8] The low-fertility scents triggered a decline in testosterone and lower interest in sex, whereas the high-fertility scents triggered a rise in testosterone along with positive ("absolutely" and "very") answers to questions such as: "Would you like to have sex right now?" and "If you were to have sex right now, how 'hot' would you be?" *Muy caliente*, indeed.

Testosterone has benefits, but as with all things evolutionary, there are trade-offs. Levels need to be high when men are confronted with mating opportunities and threats (i.e., when triggered by these social contexts, as in our challenge hypothesis study). But if men had evolved to have high levels of aggression-triggering testosterone throughout the day and into the night, they would experience more unnecessary conflicts with other men and probably take less interest in their kids[9] (and women would probably not benefit either).

This could be why testosterone is higher in the morning before men are out of bed and headed off into the world, where they could get into trouble over the course of their day. Men need testosterone to fuel their male traits, but they don't need it to be constantly elevated while they are simply going about the daily business of living.

Testosterone regulation is strategic—men, like women, have hormonal intelligence.

The Scent—and Senses—of a Woman: The Female Response to Females

Female baboons that are capable of detecting ovulation cues in other females will respond by becoming highly aggressive – to the point of deadly encounters. They specifically go after high-fertility females, perhaps because they're rivals for good-genes males. Of course, fertile female baboons are offering up more than scent cues to give themselves away (see: swollen genitalia). But there is evidence that scent on its own is an effective channel for female-to-female communication.

After confirming through our own research that men are capable of picking up women's ovulation cues, and considering the behavior of nonhuman primates, my students and I wondered if women could similarly detect female ovulation cues.[10]

Once again, as in our previous scent study, we collected scent samples of women's body odor by having our participants wear gauze pads under their arms at confirmed high- and low-fertility points, instructing them on the same scent-free protocol and "no stinky living" rules. Recruiting a panel of scent raters, however, was a bit different this time around, as we needed to find women to smell other women.

We were curious about whether a woman's experience of being in close contact with other women might sensitize them to differences across the cycle in women's scents, so we set up an information booth at a local (and very lively) lesbian and gay pride parade to recruit participants. They were a diverse group in how they identified their sexual orientation — a mix of lesbian, bisexual, and heterosexual women.

At the same time, since we had a show-and-tell mini-lab set up at the parade — with actual samples — we set about trying to do some on-the-spot research, hoping to collect data from female passersby. We did what we would eventually do in our on-campus lab: We asked women to sniff unmarked high- and low-fertility samples and rate them as "very" or "not very" pleasant, sexy, or intense.

As a scientist who likes lots of data, if I am presented with potential study participants congregating in one place (such as a crowded parade with a lot of enthusiastic women), it's tempting to engage with as many of them as possible, but... let's just say that the data were a wash, given the proliferation of "rainbow margaritas" consumed by most of our would-be scent raters. Our pop-up research lab was an effective way to recruit women for the in-lab sniff sessions, but there was no way to control that day's data for tequila and food coloring!

Back indoors, we held our rating sessions and collected results that confirmed the baboon-based predictions: The high-fertility scents were rated as more attractive (pleasant and sexy but less intense) than the low-fertility samples, a preference borne out just as it had been in men. Furthermore, the appeal of high-fertility scents held at the same level across our sexually diverse groups, suggesting that women's sexual orientation was not a factor after all.

The purpose of an ovulation cue detected by males relates to mating/reproduction, but one detected by females surely served another purpose. Going back to the female baboons, which behaved aggressively toward the estrous female, a female-to-female cue may likely serve to signal the presence of a rival. Among ancestral women, it might have served a similar purpose.

In one small study, women who smelled the high-fertility scents of other women maintained their levels of testosterone (associated, of course, with aggression). But testosterone levels declined when they smelled low-fertility scents. Perhaps they felt less threatened by the woman who was at low fertility.

There are still more female-to-female factors to take into account: What about the fertility status of the woman who detects the scent cue? If she herself is highly fertile, and therefore primed for competition, she could behave more aggressively. But she might still recognize that a fertile woman's scent cues are attractive (no sense in paying the costs of competing unless you recognize which women are particularly attractive and therefore competitive threats).

Clearly, there are more research opportunities here, including ones that would shed more light on women's social relationships. For now, however, even if we don't know their exact function, there is little doubt that female-to-female hormonal cues are real.

How Hormones Sound

Female mammals make different noises when they are in estrus. Cows moo more than usual. Elephants produce a low-frequency "estrous rumble."[11] Yellow baboons vocalize with "copulation calls."

Similarly, women's voices are altered by hormones during high fertility.[12] My colleague Greg Bryant and I collected vocal samples of close to seventy women at both low and high fertility who recorded the following simple message: "Hi, I'm a student at UCLA." At high fertility, as women approached ovulation their voices rose in pitch. Women who recorded their messages at maximum fertility (when luteinizing hormone surged to its highest point in the cycle) had the biggest rise of all. In other words, when luteinizing hormone peaked, voice pitch did as well.

In a separate study, voice raters—including both men and women—judged the higher-pitched recordings as more attractive than the low-fertility recordings,[13] perhaps because higher pitch is perceived as more feminine.

A Kiss Is Just a Kiss, But Is It Also a Cue?

Hormonal cues are detected by the senses — a woman at high fertility may look, smell, and sound more appealing than she does at other times. I know of no research on detecting cues through touch, or if the approach of ovulation might cause a woman's skin to feel softer or her hair to feel silkier. (However, the flood of pregnancy hormones is famous for triggering "that pregnancy glow," including thick, lush tresses.) But what about . . . taste? I'm not about to descend, literally, to exploring the nether regions of estrus, though a very prominent, infamously uninhibited evolutionary biologist bluntly referred to oral sex in the Q&A following a talk I gave to a packed house. (He claimed he could definitely detect ovulation when he was close to the source; all I could do was smile and say it seemed reasonable to me!) Instead, here I'm talking about standard-issue kissing — and, more specifically, about saliva, which is teeming with hormones (and bacteria, but we'll get to that).

Scent and taste are related senses; both send messages to the brain to help interpret chemical information in molecules, whether those molecules are airborne (scent) or ingested (taste). Therefore, it makes sense to move from scent to taste as another possible cue. But there isn't nearly as much conclusive research on the role of saliva as there is on the role of body odor. We do, however, know a few things about the hormonal information saliva may convey.

Drooling over You

Big-eyed but tiny mouse lemurs are prosimians, a suborder of primates. (If you don't know what this mini-mammal looks like, the too-cute character of Mort in the *Madagascar* films is a mouse lemur.) We humans are further removed from prosimians than we are from orangutans (apes and humans are simians), but mouse lemurs — like us — are still mammals, and their estrous behaviors will sound familiar.

The females in this nocturnal species have increased locomotor activity during estrus; they emit scent cues and increase their vocalizations with high-frequency trills; they also groom themselves more during this time. Their peak fertility lasts a mere two to four hours, in part due to their unusual vaginal morphology; the vaginal canal itself is open for only a few days during the cycle. (In captivity, this cycle has been observed to be as long as fifty-eight days, but it can even be as long as one hundred days.) Mouse lemurs pack a lot of fertility cues into twenty-four hours or less, and they are not subtle — because there's no time to be coy.[14]

Mouse lemurs "muzzle-rub" most frequently during estrus. As kissy-kissy as this sounds, muzzle rubbing is not kissing, and in fact is done solo, but saliva is the key ingredient. Female mouse lemurs rub their mouths (muzzles) on tree branches or, if in captivity, cage bars, licking and biting the surface. This is a form of scent marking, but it's done with estrogen-rich saliva. (Mouse lemurs also deposit scent cues through urine, another indicator of hormonal status.) Because lemurs are nocturnal, they wander in darkness and might need to rely on cues other than visual ones; this might explain the use of estrogen-infused saliva.

Saliva certainly has the potential to serve as a means of chemical communication between males and females, given that it contains steroid hormones produced by the male testes and the female ovaries. (As described earlier, we use male saliva samples to establish testosterone levels in research.) Among patas monkeys, the estrous females will actually drool as they display their backsides, complete with sexual swellings, to males for copulation. Though scientists aren't exactly sure why they do this, they could be presenting a hormonal cue. Perhaps the saliva is added insurance to make sure the male is getting the message.

Kiss-Kiss, Bang-Bang

Saliva has been largely overlooked in research as a mode of male-female communication. Given its hormone content, says Alan F. Dixson, author of the classic text *Primate Sexuality*, "the potential for saliva to act as a vehicle for chemical communication between the sexes should not be ignored."[15] Let's leave the muzzle rubbing to lemurs. One surefire, very human way for the sexes to "communicate" via saliva is by kissing. Not chaste pecks but full-on, get-in-there tongue kissing.

Hormonal cues are detected by the senses, and kissing involves the sense of taste — though that's only a part of it. The act itself can be a fully sensual experience involving visual and scent cues, sounds, and touch. (You may well close your eyes to kiss once you have confirmed your target's location, but you're unlikely to be holding your nose and keeping the other person completely at arm's length.) Kissing, done right, is the whole package.

A man can't taste for levels of reproductive hormones in a woman (unless he's some kind of sexual gourmand supertaster who has genuinely isolated the umami of estrogen). He can, however, simply *like* the taste, or find the experience arousing, in the same way he prefers the appealing scent of a high-fertility woman without knowing why. (*Mmmm...I like this person.*) In that regard, how a woman "tastes" could be interpreted as a hormonal cue. (It's worth noting here the metaphorical yet undeniable connection between food and sex, beyond the fact that a date often involves dinner. We describe members of the opposite sex as "yummy" or "delicious," as if he or she were a fresh-baked cookie. We want a taste.)

Given what we know about evolution, it seems that kissing should have some function other than pleasure and arousal — or that it served another purpose at one time in human history. Besides containing hormonal information, saliva also contains microbes — and lots of them. We know that there is a bacterial transfer during kissing — to the tune of eighty million microbes per single ten-second kiss (think make-out session — but even if your partner's kisses are half as long these days, you're still getting a healthy serving of microbes).[16]

One theory holds that the exchange of microbes — rather

than hormones — serves a genetic function in reproduction of healthy offspring. Major histocompatibility complex (MHC) genes function like human immune-system guardians, detecting pathogens and essentially kicking out invaders that might be trying to sneak in disguised as healthy cells that belong to the "self." The more complex one's MHC gene code, the better — because pathogens can't mimic the code as easily.

When two people kiss, they're exchanging saliva full of microbes that contain MHC genes. If things progress way beyond that kiss, offspring, according to this theory, are healthier and hardier (with stronger immunities) if they get two different alleles from their parents from a more complex MHC code. In other words, men who have dissimilar MHC from women bring those all-important good genes. There is some evidence to suggest that women are more MHC-dissimilar to their romantic partners than not. (In a Swiss study, women preferred the scent of T-shirts worn by men with dissimilar MHC.)[17]

As far as MHC goes, opposites seem to attract. Yet when it comes to germs, some studies show that among committed couples there are more similarities in their microbiota than not.[18] In other words, they have the same bacteria in their spit. It's unclear if this is because they exchange microbes on a regular basis; maybe they start out with different microbes and things equalize as romantic couples begin to share the same living environments. (Shared toothbrushes equal shared microbes — and, unfortunately, shared illnesses, too.)

So, kissing probably has more biological — and hormonal — significance that we've recognized. Consider that the next time you and your honey share an intimate kiss.

Undercover Ovulators: Subtle and Strategic

Women evolved to keep their fertility status all but hidden. Even though they engage in observable mate-search efforts that often coincide with estrus, they are not telegraphing the approach of ovulation and peak fertility. They are, however, emitting cues such as scent. The cues themselves aren't completely hidden from men (or from other women), but their precise *meaning* is. As I pointed out early on, men can detect a cue, but they cannot decode it with certainty.

Let's look at some possible explanations for the existence of human ovulation cues and for why ovulation itself is so concealed. But let's get something out of the way first. *Women are not signaling their ability to conceive.*

We've already established that men who are open to sexual opportunities hardly need convincing. A woman doesn't need to entice a man by exhibiting her fertility (subtly or not), and it's not clear that she would benefit from doing so. She might attract the wrong kind of attention, in fact — from unwanted males or rival, bullying females.

The human male brain has evolved so that a man does not need cues — yes, think red-pink baboon bottom one last time — for mating. While it's true that men might respond sexually to a woman's revealing clothing or a whiff of her scent, they probably evolved to hedge their reproductive bets and be interested in sex whenever their partners are receptive to sex or, better, are seeking it (better safe than sorry when it comes to opportunities to reproduce). Extended sexuality (having sex when pregnancy is not possible) and its role in bonding also come into play when sex happens: A woman expressing interest in sex outside of the fertile period can help reinforce the bond with a partner and secure his investment. But once the pleasure apparatus is in place — with the fun of sex as its driver — men might be interested in sex just because it feels good. Or maybe it's just Saturday.

Still, the ovulation cues are there — because the hormones are. Women might have evolved to conceal ovulation cues, but there might have been a limit on just how much they could do so without compromising fertility. One strategy, for example, would be to reduce the levels of hormones or reduce the hormone receptor densities in her tissues (whose functioning might include telltale signs of fertility). But this could have reduced her fertility or compromised her ability to become pregnant or to maintain a pregnancy. The "leaky cues" hypothesis proposes that ovulation cues are by-products of physiological changes in the body as it reaches peak fertility: Because of normal, cyclical hormonal fluctuations that affect multiple systems in a woman's body, cues will "leak" out, for instance, in the form of a hormone-laden scent. And men were under strong evolutionary pressure to detect these cues and find them attractive, however subtle they might be. This has resulted in a delicate coevolutionary dance between females, who have evolved over time to conceal (up to a point), and males, who have evolved to detect.

The question, of course, is, What were the advantages to women of concealing? There are several possibilities…

To Increase Paternal Investment (Good Dad Sticks Around)

One possibility involves men's paternity certainty (*My baby or the other guy's?*). If women's ovulation was easily known (or even approximated), men could just stick around during these fertile days and wander for the rest of the cycle, perhaps in search of additional mating opportunities. But if he were unable to determine when she was most likely to conceive, he would have had to stay close, mate guard, and try to impress his mate with investment — *throughout* her cycle. The more Good Dad hung around and invested in his family, protecting them and procuring resources, the easier pair-bonding and co-parenting would be. And extended sexuality perhaps shored up his paternal investment. Regular sex sealed the deal.

If Sexy Cad were on the prowl and had been able to detect female fertility, he would have dominated the mating opportunities within a group. With him in the dark, so to speak, women were able to pursue other males, including Good Dad types who offered paternal investment and big-brained babies.[19]

To Avoid Aggressive Females (More Friends, Fewer Foes)

As noted, female baboons will attack and even kill fertile females chosen by high-ranking males. (*If I can't have him, then you can't either.*) It could be that ancestral women evolved to conceal ovulation cues from other females to avoid being the targets of such aggression.

We have also seen hints that women's testosterone goes up when they detect scent cues of fertile women, even if we still don't know precisely how highly fertile women react toward one another. We also know that, according to the challenge hypothesis, testosterone rises when males encounter fertile females and simultaneously have to contend with rival males. Now, imagine if all males could sense when any female was fertile — the air would be thick with sex hormones, including testosterone, and there would be fights breaking out all over.

Given the potential impact of testosterone on both sexes, concealed ovulation may also have served as a peacekeeper that allowed humans to cooperate and create communities conducive to raising successful families.[20]

To Preserve Female Choice (She Gets What She Wants)

Finally, concealed ovulation allowed ancestral women to be selective and to thrive. A woman was able to mate shop for high-fitness genes on her timetable. In some cases she and her offspring thrived because she opted for a "mixed" mating strategy, securing good genes from the Sexy Cad but sticking with Good Dad for the long haul. And no one was the wiser.

Women have evolved to fly under the radar, hormonally, keeping their fertility on the QT and protecting themselves from unwanted aggression, male and female. But it's not just the ovulatory phase that is concealed across the cycle and across a woman's lifetime. It is *every* phase. No one can look at a woman and discern if she's menstruating or has PMS, or even tell — at least in the early stages — if she is pregnant or menopausal, and that is all to her advantage.

In a sense, female hormonal intelligence itself is hidden, but not from the person who is most empowered by its existence.

7

Maidens to Matriarchs

FOUR HUNDRED. THAT IS how many menstrual cycles most healthy women in the industrialized world will experience in total, waves of hormones month after month that will rise and fall in a fairly steady rhythm. The timing is usually quite reliable for years on end, as is its length. *Okay, I had it three weeks ago, and today is the fifteenth, which means I should get it again by Friday and be done with it by Tuesday....* But though a period itself can be fairly predictable for years on end, the bigger picture of a woman's hormonal life is marked by considerable variation.

The onset of puberty, the fertile years, and menopause can vary dramatically from woman to woman. Getting one's period at twelve, for instance, getting pregnant at thirty, and entering menopause at age fifty sounds "average" enough, a nice, neat way to plot out a lifetime of hormones. But in reality, girls can have first periods at age ten, grow into women who choose to give birth at age forty, and have periods into their mid-fifties — all of which could still be considered perfectly normal and healthy.

Compared to our ancestral sisters, modern women may experience these distinctly female life phases at vastly different ages.

(In less than fifty years, the age for first births among US women has gone from 21.4 years in 1970 to 26.3 as measured in 2014 — an inching-upward trend that shows no sign of reversing itself.[1]) Regardless of when they may happen in a woman's life, I frame these life phases in terms of "eggonomics": the biological trade-offs and shifting costs and benefits among growth, mating, parenting, and grandparenting.

As you will see, female hormonal intelligence evolved to start early — and last a lifetime.

The Price of Puberty: Who Gets to Reproduce?

There is one singular phase in a female's life when her hormones are fully on display, all but impossible to disguise, and that is when puberty is in full swing. Boys are not exempt from the hormonal explosions that seem to coincide with the middle school years (cruel timing, indeed), but for girls, the changes seem particularly obvious.

Talk to the mother of any girl entering puberty and she'll describe a child who still plays with toys, watches cartoons, and likes to snuggle, but who also longs for parental independence and privacy. Girls who take an interest in makeup and clothing seem to transform in appearance overnight, even if the American Girl dollhouse still takes up a corner of the bedroom. Those who have little regard for lip gloss or the latest styles still can't escape the estrogen-related changes in their bodies: breasts, hips, and facial features that become fuller, more defined, and mature.

At peak fertility within the cycle, women may feel more competitive and aggressive toward other women, and it's tempting to draw a line from this behavior back to the stereotypical "mean girl" antics of some girls and young women. But some of this tension, of course, is simply environmental. It's hard for a still-dependent young person to escape the hothouse atmospheres of middle and high school, complete with their social pecking orders (and increasing academic pressures — reminders of the real world to come).

But some behaviors truly are hormonal.

Girlhood: The Challenge of Acting Your Age

Girls are capable of becoming interested in sex well before they are ovulating, are having periods, and are fully fertile. But being "boy crazy" doesn't mean that a girl is headed for a teenage pregnancy or even for sexual intercourse itself. (In fact, teen pregnancies have been steadily declining in the United States for the last few decades.[2])

"Subfertile" (not-yet-ovulating) girls can still be interested in boys. This supports the notion that sexual desire in humans is not just about reproduction; it is also about forming relationships. Girls who chase boys or who like to be chased could simply be sampling the possibilities. By "playing house," girls might be learning various skills in the context of a domestic relationship, including figuring out who might be a good mate and partner in parenting. Knowing that, if you are the parent of a young child, you may never watch a playdate in quite the same way. *It's Kiran's turn to be the daddy and his job is as a TV reporter. Mollie is the baby who gets nervous sometimes because she misses her mommy. I'll be the mommy, and I am a veterinarian and an astronaut. We have nine dogs and eat ice cream for dinner.*

The age of menarche — the first menstrual period — varies widely, and it's one reason why two eleven-year-old girls at the same sleepover may look like they are years apart. There are numerous reasons why hormones kick in on different schedules, including nutrition (and obesity, linked to earlier menarche), environmental influences, race, heredity, and more. (I'll address why certain chemicals and toxins can also cause variations in a hormonal cycle in Chapter 8.)

In ancestral times, when physical and social conditions could be unstable, it is likely that nutrition — enough food versus a scarcity — played a major role in a girl's sexual maturation. Malnutrition could have resulted in later menarche and fewer fit offspring during lean times.

But the reverse eggonomics outcome was also possible. When times were "good" and conditions allowed for more of a feast than a famine, young, healthy females could have matured earlier to produce more fit offspring for a longer stretch of time. Interestingly, female wasps exhibit this use-it-or-lose-it strategy with respect to fertility and offspring-friendly conditions. Wasps generally invest time looking for just the right spot to lay their eggs, but in experiments that manipulate their life expectancy (a daylight pattern associated with the fall season, when lives are shorter than in spring), they stop looking and they lay their eggs quickly.[3]

Girls may form some of their deepest friendships with one another as they go through puberty, but it can be a time of social conflicts among same-age females as well, in part because girls are maturing in front of one another at such different rates — and in front of males who might take notice. Girls who mature early may get unwanted attention from boys (and men) and from other females as well. And those who are late to develop may hit their own rough patches. Fortunately, puberty doesn't last forever, and eventually the initial pubescent rush of hormones will settle into a more steady rhythm.

Mother-Daughter Conflict

Puberty is also peak time for a girl's conflicts with her parents, particularly with her mother. A mother and her daughter could have estrogen levels at opposite ends of the spectrum (one beginning to wane, one beginning to surge), which may nudge moods and behaviors in opposite directions as well, though the sentiments each expresses in the heat of the moment may sound like they're coming from an echo chamber. *You don't care about me. Why are you so mean to me? I can't wait till we're not living under the same roof.*

Modern as it may seem, this classic mother-daughter head butting may also be the remnants of an ancestral reproductive conflict. As soon as she was old enough, an ancestral girl likely became a valued helper at the nest, caring for offspring and taking on domestic duties, particularly if her mother's reproductive years were not yet over.

In some nonhuman primates, younger females who remain with their mothers or other high-ranking females are likely to have delayed sexual maturation — perhaps so that they remain in the group as subordinate helpers.[4] But there could be another reason (beyond the free babysitting) that an ancestral girl who stuck close to her mother remained relatively subfertile: If her childhood environment was fairly stable — good conditions, plentiful resources — there was no pressure for her to reproduce, and she could take more time moving through her development and building up her body.

Eventually, as a daughter grew older and reached puberty, she would experience her first estrous desires and feel pulled to engage in her own mate search and produce offspring of her own, just as her mother had done before her. There may have been ancestral "maiden aunties" who, for whatever reason, set aside their own reproductive opportunities to stay behind and care for the matriarchs and their children. But more likely, girls evolved to become independent from their mothers and move on.

Puberty may trigger tears and frustration, fray the bonds of friendship, and test the patience of even the most understanding parent. But the hormones involved in this female passage helped to push ancestral girls from the nest once they were ready — and they're still doing it today.

The Cost of Pregnancy: Mating Mind versus Mommy Mind

It may be easier to get pregnant when you're twenty-five than when you're thirty-six, but for most career-bound women today, it's wildly inconvenient to do so. There is a very real conflict between peak career-building years and peak fertile years, as too many women in (and out) of the workforce will attest. Modern life steers us to delay the decision to produce offspring until we ourselves are more prepared in a variety of areas, from physical health to material resources, and, typically, until we have a mate to help at the nest.

In postponing motherhood, we are perhaps very much like our ancestral sisters, who were seeking more than just a male to fertilize an egg. Those who waited until they had a better chance of carrying a pregnancy to term had offspring with what the literature calls "greater fetal outcome" — simply put, healthy babies who survived and thrived, and who would themselves go on to reproduce.

Older, Wiser, and Still Fertile

Early-modern human pregnancies probably occurred most often among women in their twenties, when they would have several rounds of conception, childbirth, breastfeeding, and renewed ovulation. This actually goes against what we may think, which is that given the lower life expectancies of our ancestors (hunter-gatherers who died in their forties), women became pregnant as soon as they were ovulating and fertile, at very young ages.[5] In fact, the average age for first birth among humans was closer to nineteen years of age, not thirteen to fourteen (the age at which chimpanzees and bonobos first give birth; for gorillas it is ten).

Very young females capable of becoming pregnant would have been at a disadvantage in terms of their health, as their bodies were still growing. Immune system development, bone growth, and brain development during puberty, for instance, were extremely important for long-term survival. Pregnant girls would have naturally shared their body's resources with a developing fetus, but they were still maturing physically; pregnancy would therefore lead to a competition for resources between parent and offspring.

If a girl becomes pregnant before her pelvic bones are fully grown and mature, for instance, the fetus will still take whatever nutrients it can from her body for its own skeletal formation. The girl's pelvic bones therefore won't grow normally, and narrow hips do not make for easy childbirth. Because of the competing demands of a girl's own growth and that of the fetus, there is the risk of a variety of complications in the offspring associated with lower access to nutrients, such as prematurity, low birth weight, and even stillbirth. This type of maternal-fetal conflict would have had a high cost for both mother and child.

Young females might also have had a disadvantage if older

females viewed them as competitors for fit males. Sisterhood isn't always powerful if mates and resources are limited. Furthermore, older females would naturally have had more experience than their little sisters at reading their physical and social environments — they would have been more skilled at assessing their reproductive environment, including which mates are good ones. To take it back to our fishy cousins who are part of the origin of estrus, female guppies that view a male in the act of mating find him more attractive than males they do not observe in the act.[6] It is as if the females are looking for proof — from *other* females — that the male is desirable and capable, the guppy version of looking for the Sexy Cad with high-fitness genes. This is a form of social learning, just as is discovering where the food is and who the predators are, and humans, of course, also engage in this important task in order to survive and thrive. But these lessons come with time. The older you are, the wiser you'll be.

Had an ancestral female chosen a mate at a very young age, she could have been bound to him for a long time because of their shared children — perhaps for most, if not all, of her reproductive life. Women and men might often have had few choices, given small group size (and no school campuses or dating apps). But it is likely that they also had choices much of the time. Otherwise, it is hard to understand why we appear to have evolved criteria for mate choice that tap into reproductive potential, resource acquisition potential, and good genetic material. But choosing too early, before learning about the social environment and the mating options, would probably not have served a woman well, unless she had managed by chance to snag the rare Sexy Cad / Good Dad total package.

In terms of a female's age and environment, becoming pregnant when the conditions were right offered reproductive advantages for ancestral women. And in the modern era, waiting until the time feels right still makes sense.

How to Succeed in Baby Making without Really Trying

I postponed motherhood till my mid-thirties for personal and professional reasons, and when I felt ready, I didn't just have baby fever—I had full-on baby panic. One heart-stopping statistic from the American Society for Reproductive Medicine states that a woman in her thirties has about a 20 percent chance of conceiving each month; by the time she is forty, her chances decline to a mere 5 percent. On a particularly bad day, I remember blurting out, "I do not want to become an evolutionary dead end!"

I was somewhere in between those crazy-making percentages by the time I went to Dr. Bob, a cowboy-boot-wearing fertility specialist who was diminutive in stature but had a supersize personality, announcing early on—elbowing his nurse in the ribs while he did so—that he could get me "knocked up with twins." (He did.) For me, using in vitro fertilization was the right choice, and I was rewarded with two beautiful children who transformed my life. But there is no question that pregnancy—particularly after age thirty-five—is a topic fraught with worry, much of it unnecessary because it stems from widely circulated but flawed information. (I must have met a dozen moms of twins who had an "oops" after IVF, which underscores the point that the baby panic might be overblown or perhaps a fertility suppressant in itself.)

The Internet will tell you that one in three women between the ages of thirty-six and thirty-nine will *not* get pregnant after a full year of trying—and the standard medical advice for women in this age range is to "seek professional help" if they've tried for six months with no luck, which is exactly what I did.

But that information is based on French birth records from the seventeenth and eighteenth centuries. Data roughly as old as the Palace of Versailles were sending droves of women to find their own Dr. Bob. As Jean Twenge, author of *The Impatient Woman's Guide to Getting Pregnant*,[7] who called out that statistic, points out, there were also no antibiotics, electricity, or fertility treatments back then. The state of women's reproductive healthcare is considerably better today than it was during the time of the Sun King.

Researchers have looked at current data and found that women between the ages of thirty-six and thirty-nine have an 82 percent chance of conceiving within a year. As for younger women, their chances are better—but not by much. Those between the ages of twenty-seven and thirty-four have an 86 percent chance. These healthy statistics hardly point to an infertility crisis.

One reason, perhaps, that we remain quick to jump to conclusions about how hard it is to get pregnant is that we focus too much on age alone rather than taking in the whole story our cycles tell. Pregnancy, or avoiding pregnancy, may be the goal. But either way, if you understand how your *own* cycle functions, you can use your hormonal intelligence to get what you want.

Eggonomics 101: Pregnancy Brain

The physical and psychological transformations a woman experiences during pregnancy are well-known. Some very specific hormonal changes, such as the rise in the hormone prolactin, which allows for milk production and breastfeeding, are temporary. But others may be less fleeting — perhaps permanently changing the female brain — including changes to a woman's cognitive function.

When I was in the early stages of pregnancy, I attended an important meeting at my university with the dean (my boss) and vice-chancellor (the boss of bosses). The proceedings were not insignificant, as I had been appointed to an important post and was reporting back to them on my work. But I wanted nothing more than to crawl under the sprawling conference room table, ball up my coat into a pillow, and take a nap. In my first trimester, I could smell everything — times twenty. For weeks, I alone detected the scent of something I could only describe as "suspicious" (and abhorrent), which I swore was in the kitchen somewhere. *Where was it?* Finally, I found a tiny, well-covered, but decaying can of Fancy Feast pushed to the back of the fridge. I was not much of a barfer, but that day I tossed my cookies in the kitchen sink. At around the same time, I also noticed people's faces in a different way — and certain people, whom I'd previously found tolerable if quirky, creeped me out.

All these disparate sensations stemmed from the same source: During pregnancy, hormonal intelligence works to tune up the mind so that the body can better handle a new set of challenges — including protecting the immune system and avoiding danger. Progesterone, for example, which is sustained at a high level during pregnancy (as opposed to rising and falling during a regular cycle), causes women to love a good nap and steers them away from things they find disgusting (like rotting cat food) or dangerous (like threatening individuals).

My colleague Dan Fessler, whose study on high-progesterone women finding images of unhealthy males to be particularly unappealing is highlighted in Chapter 3, wondered about what he terms "elevated disgust sensitivity" in the first trimester. In his study he asked pregnant women which disgusting things (cockroaches, exposed organs, worms, mucus, wounds, toilets, bestiality...) triggered the highest levels of overall aversion when they were in their first trimester — when the fetus was most vulnerable to infection and the possibility of miscarriage was highest.[8] The greatest aversions involved risky food choices such as drinking spoiled milk. This makes sense when you consider that ingesting toxic bacteria from tainted foods can result in salmonella poisoning or listeriosis, causing fetal development problems or miscarriage.

The first trimester is when morning sickness is most likely to occur — and it is also when the pregnancy hormone human chorionic gonadotropin (hCG) is at its highest. Though researchers aren't sure if that hormone is responsible for nausea and vomiting (and Fessler doesn't point to hCG in his study), many have found a strong link between its presence and morning sickness.[9] Even if hCG has just a supporting role in stopping a pregnant woman from ingesting harmful pathogens that could hurt a developing fetus, it is part of a larger process that seems to confer a protective benefit to mother and child alike.

Besides being able to detect and avoid pathogens in my own refrigerator when I was pregnant, I also experienced something else that had never been an issue for me before — I felt like a fog had descended over my brain.

Maybe I was just needing a nap, but I was spacing out so often that I wondered if I was suffering from the infamous "pregnancy brain" — an inability to pay attention and to recall things. I'd always thought this was an outdated myth. As an evolutionary psychologist — and a woman — I could hardly believe that natural selection would make us weak-minded just as we were preparing to reproduce and needing reserves of physical and mental strength.

But here is a cost-benefit lesson from eggonomics: Becoming pregnant is one of the most metabolically challenging things a woman's body will ever do, and there will be energetic trade-offs. Research has shown that there is a measurable if slight decline in memory and some other aspects of cognitive function during pregnancy (and after birth), but it is happening because the body is diverting resources to fetal health.

(Mommy) Mind over (Gray) Matter

Psychologist and researcher Laura Glynn has studied the impact of hormones on the brain during pregnancy, and her work reminds us that the levels of hormones flooding a woman's system to maintain a pregnancy are vastly higher than they are during a normal monthly (nonpregnant) cycle.[10] Estrogen, for instance, is thirty times higher. There is so much change taking place in the body that it is understandable that there would be an impact on the brain as well, including on its function and its actual structure.

Glynn, who points out that much of what we know about the brain during pregnancy is based on rodent studies, was among the first to study humans. She recruited more than 250 pregnant women to come to her lab and analyzed their cognitive function during and after pregnancy to learn how this particular hormonal phase in a woman's life might give rise to "mommy mind." Glynn also wanted to know if there was a difference between how women might be affected during a first pregnancy versus a second (or third, fourth, etc.), so she included both "primiparous" (first-time) and "multiparous" (having had more than one birth) moms in her research. (She also looked at the brain function of nonpregnant women as a comparison.)

In Glynn's study, which used a classic verbal recall task, a woman's memory declined as her pregnancy progressed, starting at about the fourteenth week (or early in the second trimester); up to this point, there was no difference between a pregnant woman's performance and that of a nonpregnant woman. Starting at about the twenty-ninth week, the differences between the two groups were even more pronounced. Higher levels of estradiol in early pregnancy were associated with poor performance on the recall task. Three months postpartum, the effects on memory were still evident, and the multiparous mothers performed worse than the primiparous moms. (That last finding may not come as a surprise to any mother of an infant with an older sibling or two.)

No, the baby doesn't literally eat your brain, even if it feels that way. But there is evidence that a woman's brain plasticity is somehow altered by pregnancy. (In a recent study of pregnant women's brain scans, gray matter appeared to be reduced in volume — in particular, in the part of the brain that dealt with social cognition.[11]) Memory, it seems, is indeed affected when the brain is rewired by the hormonal surges of pregnancy; in fact, more than 80 percent of new mothers report having memory problems, even after controlling for sleep deprivation, stress, and depression.[12]

Most research, however, shows that this memory loss is concentrated in the area of verbal recall — for example, remembering words previously supplied to you — not really big stuff like remembering to take the baby out of the shopping cart and put him in the car before you drive out of the parking lot. During pregnancy, as a woman's body prepares for birth and motherhood, there seems to be an eggonomic cost to this hormonal transformation: a slightly foggy brain. But as you're about to see, there is also an upside to "mommy mind." In exchange for a bit of memory, mothers enjoy heightened sensitivity, surveillance, empathy, and intuition, all of which make them better parents.

Feathering the Nest

When I was pregnant, I wanted to nap — and I wanted to get organized! Twins were coming! We needed cribs, car seats, clothes, diapers, wipes, more diapers, hypoallergenic-organic "cleansing water," herbal baby remedies, tiny nail clippers, special strollers to snap those car seats into for the eight-foot trip from the car to the front door.... I obsessed over the design of the nursery, making it boy and girl "appropriate" (aqua!), soft, cuddly, perfect for these precious creatures about to inhabit the planet. Just the right bouncy chair, a high-tech swing with womb sounds, baby board books — never too early for reading... I learned what I could about basic infant care, how to swaddle to soothe, crying, feeding, pooping, sleeping.... Plus, babyproofing! A must! (I was determined to install child-safety locks on all the kitchen cabinets before my twins were born, though they would be sedentary butterballs, as well as bald, for their first year of life.)

On top of it all, I felt an overwhelming urge to clean.

Why would a hugely pregnant woman exchange the urge to get some much-needed rest for the burning need to run a vacuum cleaner, assemble furniture, and reorganize the linen closet? The nesting impulse is real, but you may wonder if it really serves a function. A working mother-to-be may want to clear every scrap of paper off her desk and provide copious instructions for colleagues in anticipation of her maternity leave. There is good reason for that kind of advance preparation. But in addition to putting her office in order, she'll probably want to bake and freeze twenty lasagnas and regrout the shower. No, she is not possessed by Martha Stewart; there is something else at work here. Along with the estrous desires that can result in offspring, humans and animals share a powerful and purposeful nesting instinct.

Across species, the most vulnerable time of life is at birth and in the minutes, hours, and days that follow. The threat of death — from disease, from injury, or from a predator — is high, not only to the newborn but to the mother as well. One way to reduce this risk is for a pregnant female to prepare a nesting spot that will provide security and hygiene. Birds do it, bees do it (but those wasps don't when they really need to lay their eggs ASAP — see the discussion earlier in this chapter), and we humans do it too — especially in the third trimester of pregnancy.

In one study on the "nesting psychology" of women in all three trimesters of pregnancy (as well as nonpregnant women), those in the third trimester were more likely to report nesting behaviors than other women — specifically organizing, throwing things out, and creating new spaces.[13]

In the first trimester, when the risk of miscarriage is highest, women, as discussed, have been shown to have high levels of sensitivity and aversion to factors linked to disease. Several months later, it seems these gate-keeping behaviors reemerge transformed. Researchers surmise that the urge to hang new curtains, paint the house, or wash the floors is related to the adaptive behavior of reducing pathogens, or at least dust bunnies.

Women also report a burst of energy in the third trimester, as if they are racing against the clock to prepare the nest — and they are. This might remind you of another hormonal phase involving restless energy: when women at high fertility move more and eat less. It seems that women in their third trimester do an eggonomic trade-off of their own, when they move more and nap less — even though sleep will soon become a precious commodity.

The Mama-Bear Effect

When I was pregnant and staying at our cabin in the mountains, one of the locals — a salt-of-the-earth woman who hand-fed the bears — suddenly seemed particularly troubling to me. For years, she would routinely appear on our deck with no warning and peer into the house. She had no boundaries and no filter, but my family had grown used to her seemingly harmless antics. Yet now I felt a fierce urge to run this person off the property. One day she materialized as usual, looking in the windows, but I grew wary, and my dog — who was usually friendly — actually snarled at her, as if he sensed my trepidation. When she approached me later at a neighbor's Fourth of July party to complain about the dog, I — very uncharacteristically — raised my voice and gave her a chewing out of my own, as if I were protecting my unborn twins (or their current surrogate, the dog). It turns out, that's probably what I was doing.

I was midway through my pregnancy when this happened, but the feeling hinted at what was to come. Pregnant women develop protective mechanisms to avoid risk — not just from pathogens but from predators, too. Studies show that during pregnancy, women exhibit sharper facial recognition skills, particularly for male faces they deem threatening.[14] (It would be interesting to find out if pregnant women are better than others at picking criminals out of a lineup.) While "pregnancy brain" may have a negative impact on memory, it seems some aspects of memory — or at least our cognitive performance — could be enhanced.

When my children were tiny babies, I felt flashes of fierceness as I pushed their double stroller down the sidewalk, ready to throw myself between them and whatever might cause them harm. Before they were born, I would smile and nod at strangers walking their dogs — I am a devoted dog person — and often stop for a chat and pat. Now when I saw a dog on the street, the thought that ran through my head was, I will break you in half! It didn't matter that the animal was uninterested in eating my children. I felt ready to protect and defend my offspring to the death, just as mothers have done for millennia.

The hormonal changes of pregnancy and childbirth are famously transformative, and the impact goes beyond the physical. The Mama-Bear Effect, which I experienced myself, is an obvious reference to the old idea that it's unwise to get between a mother bear and her cubs. Jennifer Hahn-Holbrook, a psychology researcher (and my former postdoctoral student) has done extensive work in the field of pregnancy and postpartum behavior, including this tendency toward "maternal defense." She reframes the Mama-Bear Effect this way: Never get between a lactating mother and her offspring. Her studies have shown that lactating women, who have high levels of the hormones prolactin and oxytocin, are quicker than other females to become aggressive in the face of a threat.[15]

In rodent studies, scientists had observed that it was not very easy to stress out a lactating rat, even though they tried: running them through mazes, forcing them to swim, setting up encounters with predators or other threats. Still, the rat mothers remained calm, exhibiting few changes in their stress hormone levels and cardiovascular function. Study after study showed that lactation seemed to confer a stress-reduction benefit. But at the same time, the rat mothers were more aggressive and faster to become defensive when threatened than nonlactating rats.

Jennifer wondered whether humans behaved in the same way. She designed a study with three groups of women: nursing mothers, formula-feeding mothers, and women without children. Each participant was initially confronted with an irritating female research assistant posing as another study participant. (Moms also brought their babies to the lab, so technically their offspring were within the range of the threat and not safely at home.) The assistant would loudly chew gum, check her cell phone, not make eye contact, and generally be obnoxious. Each participant was told she'd be playing a computer game against her annoying opponent that measured her reaction time to a particular task; whoever "won" would be allowed to send out a blast of sound to the "loser."

The participants had baseline levels of blood pressure, heart rate, and other stress markers taken. They were then sent into separate rooms to play the game. In actuality their "opponent" was a computer program designed to deliver loud blasts of sound to the participants (apparent aggression from their apparent opponent), to see how they would respond. The participant assumed she was getting "blasted" by the gum snapper, and she could respond by returning her own burst of sound. After playing the game once so that they were familiar with the setup, participants took a break so that mothers could breast- or bottle-feed their babies; non-mothers read a magazine. Then they played again. Afterward, the researchers took measures of blood pressure and other markers a second time.

Jen's results yielded this remarkable finding: Breastfeeding mothers delivered significantly longer, louder blasts of sound compared to bottle-feeding mothers and non-mothers. While all participants showed signs of stress the first time they played the game, when they played after the break (after women fed their babies), breastfeeding mothers had the lowest increase in blood pressure and other stress markers. As it turns out, the lactating women who delivered the loudest blasts to the confederate were those who had the lowest blood pressure. Cool. Calm. And *do not* mess with me. These results are consistent with the rodent studies: lower levels of stress, but higher levels of aggression when needed. Jennifer's work highlights the hormonal impact of breastfeeding, which allows mothers to become assertive while still remaining calm — mama bear in action.

A Mother Just Knows

The hormones that accompany so-called pregnancy brain (and the mommy mind that remains after the birth of a child) are responsible for a range of maternal superpowers. Oxytocin and prolactin contribute to the Mama-Bear Effect. High levels of estrogen at peak fertility, as we've seen, help women to avoid threatening situations and individuals. If pregnancy occurs, the protective benefits of several hormones, including progesterone, assist women in guarding a developing fetus, whether by avoiding iffy food choices or by staying away from sketchy situations and individuals.

Once offspring arrive, mothers get even better at determining what might be dangerous and what is safe. Ask any mom — including those with grown children who may live away from home — and most will agree: When it comes to the well-being of her child, she *just knows* when something isn't quite right. My colleagues and I have studied this phenomenon, and once again, hormonal intelligence plays a major role, even if the threats are subtle.[16]

Earlier in human history, the death of an infant from disease, accident, or attack by a predator was much more common than it is today, and therefore being psychologically attuned to these risks made evolutionary sense. Though infant mortality in our modern world is much lower, many ancient aspects of maternal sensitivity appear to remain intact. For instance, in experiments using recorded infant cries (a mixed sampling of unfamiliar infants' cries along with those of women's own infants), mothers successfully identified cries from their own babies, starting in the early days and weeks of life. (Dads were less successful; mothers may be more prepared to swing into action when they hear their infants in distress.[17])

Mothers were also quite prone to "postpartum precautionary preoccupation." Translation: when you worry about worst-case scenarios involving your helpless infant, as I did. From burning buildings to rabid dogs to home invasions, I entertained every child-in-peril-movie-worthy scenario, as if I were practicing for what I'd do in the event of a real crisis. Sample precautionary behavior: when you check on them at night every two hours to make sure they're breathing. What new parent hasn't done that? One major source of anxiety for parents (which lasts beyond the newborn phase) is "stranger danger," and here's an interesting finding: Just as women at high levels of fertility overestimate the size of males they perceive as threatening (see "The Dangerous Stranger" in Chapter 5), in studies both mothers and fathers estimate unfamiliar males to be larger and more muscular than nonparents do.[18]

So, fathers also engage in precautionary behavior, researchers have confirmed, but moms have the lock on worrying, even if we've updated our concerns by about 250,000 years. We are no longer worried that hungry animals are prowling around our shelter at two a.m. and salivating over our young. But we are worried that the air-conditioning may be too cold in the baby's room. *Did she kick her blanket off? Wait, are there too many blankets? Did he roll onto his stomach?* We fret over hygiene. Obsessive babyproofing is the twenty-first-century equivalent of trying to prevent one's offspring from eating a toxic berry. Parents have been saying "Get that out of your mouth!" since humans had language.

This is what reproductive hormones do. They rewire the maternal brain to make us better multitaskers, allowing us to care for our offspring in the most effective ways possible. Much to the chagrin of every kid who has ever tried to get away with something, a mother really does seem to grow eyes in the back of her head.

Bye-Bye, Baby . . . Hello, Kitty

The idea of sexually active and alluring "older" women preying upon younger men—hence the term "cougar"—predates *The Graduate*, in which Dustin Hoffman was a mere cub who fell into the playful, well-manicured claws of sleek Anne Bancroft. You'll find such females lustily paired up with much younger men throughout history (Catherine the Great) and on TMZ (Madonna, Mariah, Demi . . .). Characters like *Sex and the City*'s Samantha have been a pop-culture staple at least since ancient Greece, when Queen Phaedra developed designs on her own stepson. The band Van Halen added to the oeuvre with "Hot for Teacher," and there is the once-popular *Cougar Town*, a TV series featuring a former *Friends* cast member (who was hardly past her prime). Fortunately, the cougar-friendly term MILF seems to have scuttled into some dark corner of the Internet.

Is this all just male fantasy, whether she looks like Courteney Cox or septuagenarian Ruth Gordon in *Harold and Maude*? Or do estrous desires actually push women who are moving past their reproductive years (including perimenopause) to seek out younger men . . . or to just want more sex in general, age be damned? Such questions have been explored by behavioral scientists in recent years, in part because the pop-culture message was so prevalent. Their scientific curiosity was real, but the cougar, perhaps, is not; there is no definitive proof that women of a certain age consistently prefer younger men.

There are two intertwined issues here that deserve separate review—an uptick in sexual desire (or the freedom to indulge it), and age. Some studies do indicate that women may have an increased sex drive as their fertility declines, though this is demonstrated largely through self-reported data about desire, not measurements of how frequently women are engaging in sexual behaviors with a partner.[19] If a woman has a very low chance of becoming pregnant due to her age, there is no biological reason to engage in sex for reproductive purposes—so sex just for fun with the pleasure machinery already in place goes to the top of the list. From an evolutionary perspective, too, abandoning the offspring goal makes sense, though it's hard to say why we would have evolved to have sex in later years for pleasure only—perhaps the intimacy we experience is a continuation of pair-bonding. (Some have argued that if a woman is still even slightly fertile, this is nature's way of sending her on one last reconnaissance mission for fit genes—the late-in-life baby. But there are plenty of reproductive perils there, as well.)

As for age, when men choose younger women it rarely attracts attention. Older women, who are forever reminded that with age comes the loss of youth and beauty, cause a scandal. Younger men may like older women who are more sexually liberated and experienced than they are, a notion that undoubtedly inspired the Van Halen boys to pen their naughty mash note to Teacher. A guy may think: She's not interested in marriage and kids; she's got fewer strings attached—and therefore fewer ways to tie me down (unless she's into that . . .). And in reality sexually active older women *are* more experienced and probably more relaxed about the whole enterprise.

Whether or not they've given birth and raised offspring, women approaching and entering menopause have less biological use for "mommy mind." Shifting over to more "mating mind" is harmless (and perhaps functional for bonding), if not more enjoyable, as it's sex without the goal of reproduction or the worry of unwanted pregnancy.

Roar.

The Value of Menopause: Closing the (Fertile) Window, Opening a Door

Though early humans had a shorter life span than we have now,[20] ancestral women, who probably had most of their children in their twenties, were capable of living long enough to experience the final phase in the hormonal cycle: menopause. An extended post-fertility life span is exceedingly rare — few other mammals have this, including our closest primate cousins. Though nonhuman primates are capable of giving birth into their forties, the maximum life span of a female gorilla, for instance, is about fifty-four years (in captivity); in contrast, a woman can live for decades after her last fertile period, into her eighties or beyond.[21]

At some point, human ovulation comes to a halt, and the hormonal cycle — including estrus — is transformed with age. But menopause is hardly a sign of a woman's body failing, wearing out, or "drying up," as society often views the aging female. On the contrary, this is a new chapter of life rich with possibility and freedom — and there is no shortage of women who rack up astounding achievements in their sixties, seventies, and eighties.

Menopause may also be a valuable eggonomic phase for those mothers who exchange caring for their own offspring with caring for grandchildren. Of course, not all postmenopausal women become engaged in raising the next generation. Mothers and their adult children may live far apart or be estranged. Perhaps there are no grandchildren at all.

But at some point in human history, there were reasons why women evolved to have this final burst of hormonal intelligence, one that helps them shift toward grandparenting at the right time, and one that hardly any other animals experience.

Whale Watching: What Orcas Know about the Change of Life

Most species — including insects — are capable of reproducing right up until they die. Elephants can give birth to offspring into their sixties, giving new meaning to the term "older mom," and Antarctic fin whales can do so into their eighties — but like a vast array of females in most species, not long after their last birth, these animals will take their last breath. At least they are long-lived. Pity the poor mayfly, which infamously lives only to mate, reproduce, and die within twenty-four hours. Salmon spawn and expire soon afterward.

And of course, no one wants to tell the kids what happened to the remains of Charlotte the spider (of *Charlotte's Web*) after she spun her miraculous web to save Wilbur the pig, then had her babies and gracefully passed. In reality, Charlotte's cheerful, tiny offspring, which appear at the end of the story, would have devoured their mother's dead body as their first meal, as newly hatched spiders are known to do. If any mother has earned the right to have a martyr complex, it's Charlotte. Extended grandmotherhood, then, is out of reach for most creatures great and small. But not for killer whales, which — like us — have menopause.

Scientists have long observed that female killer whales live for decades after their last go-round at mating and giving birth. But they don't just go off and fade away, living alone or among other aging females. Instead, these older orcas stay close to their offspring and help out the next generation, particularly when it comes to finding food sources.

Killer whales organize themselves in matrilineal groups (that is, related through maternal descent). There could be anywhere from one to four generations of whales in a single matriline, all descended through the oldest living female: grandmothers, daughters, and grandchildren — even great-grandchildren. When they become sexually mature, females will go off to mate, but then they generally return to their matriline, where their calves will become a part of the group led by the matriarch. Males do not live with their children — they stay with their mothers.

As with ancestral women, whale mothers that are still fertile and their sexually mature daughters are capable of experiencing estrus and giving birth within the same time frame. But — as with humans — there are reasons why such "co-breeding" is not hormonally intelligent. When a whale mother gives birth within two years of the birth of her daughter's offspring, statistics show that the older female's calf is 1.7 times more likely to die.[22]

Researchers believe this is probably due to a finite food supply. Killer whales have a communal approach to food; they hunt together and they eat together, sharing resources. A younger female — which prioritizes feeding her own young to ensure their survival — is not inclined to share food with another female's offspring, including those of her own mother. The daughter will invest more time and effort in feeding and caring for her own offspring, and she will essentially outdo the mother in this regard as she naturally becomes stronger and faster than her parent. Older females understand that they can't compete with younger ones, and they — wisely, it seems — turn their attention to helping their daughters and other females in the group raise the next generation. (Among ancestral women, there were also costs when it came to co-breeding, but in the case of humans, it was in part because women *didn't* live in female-dominated — and female-supportive — clans, as killer whales do.)

Consider that a young woman would leave her family group once she was sexually mature and her estrous desires nudged her to secure a mate and produce her own offspring. Initially she would be living among non-family, but she could, with time, produce enough offspring so that she was related to most of those around her. Once she was surrounded by a set of children who were themselves reproducing, it made sense at some point in her life to shift from investing her time in producing new offspring (whom she might not live to see through to reproductive maturity) to instead investing in her grand-offspring.

Either way, she'd be investing in her genetically related kin, though her grandchildren would share a quarter of her genes, whereas her own children shared 50 percent. It's another trade-off: Have your own offspring, which requires competing for resources and a risk that you will not live long enough to see them reach reproductive maturity; or divert your time and effort toward caring for grand-offspring. Later in life, choosing grandmothering makes sense.

And so a female began to outlive her own fertility as she moved into menopause. She no longer needed to reproduce by having her own offspring, but her help at the nest still provided a benefit in terms of her overall fitness.

One way that "grandmother" whales maintain their dominance and value to the group is by being the experts at finding food. Offspring — especially adult males (see "The High Cost of Being a Mama's Boy," below) — are dependent on their mothers for a long time, and mother knows best when it comes to locating salmon, the major food source, as well as other prey. Older females know how to hunt, and they lead the way, passing their knowledge on to younger females. A twenty-first-century human grandma gets extra points for sending out cookie care packages, but when a killer whale locates food for her children and grandchildren, she is providing a multigenerational lifeline and helping her descendants avoid the evolutionary dead end. (But I speak for grandchildren everywhere when I say: Please keep sending the snickerdoodles.)

The High Cost of Being a Mama's Boy

Older female whales seem to play a unique role in ensuring the survival of their male offspring. Killer whale calves remain dependent on their food-finding mothers for years, though sexually mature females will leave their matriline temporarily, when it is time to mate. Males are known to hang around their moms for the long haul, even after the mothers enter menopause. Unfortunately, though, this whale equivalent of living in the basement after college does not bode well for the longevity of the sons.

Because males are so reliant on their mothers,[23] their mortality rate spikes upon their death. Compared to a female, a male is three times as likely to die a year after his mother's death if he is less than thirty years of age. His odds worsen with age—eight times more likely if he's over thirty.[24] (Female orcas, like humans, stop reproducing in their forties and can live for many decades after that.) But the reverse is true as well: The longer a mother whale lives, particularly beyond her reproductive years, the longer her son seems to live, too—and therefore the longer he'll be able to mate and father offspring. (He may be living in his childhood bedroom, but he will have sleepovers with estrous females from other pods.)

When it comes to saving the whales—through enhancing the longevity of males—long-lived mothers provide a benefit, and therefore so does menopause.

The Gift of Grandmotherhood

Most nonhuman primates don't experience menopause, as I mentioned earlier. Although humans and chimps can give birth into their forties, chimps in the wild, for instance, become weak and frail by the time they hit that age range and rarely make it to fifty or beyond. Bonobos, gorillas, and orangutans are longer-lived.

They can live into their early, mid-, and late fifties, respectively, but in their natural environments, they rarely go beyond those ages.[25]

Of course, comparing the life spans of primates in the wild to those of humans living with modern conveniences, including medicine and a reliable food supply, may seem like comparing apples to oranges (or monkeys to people). But when researchers looked at the life expectancies of today's remaining hunter-gatherer tribes — such as the Hadza people of East Africa, who live much the same way they did thousands of years ago — they found that about 75 percent of the population lived past age forty-five, into their fifties and sixties. While it's true that their life expectancies are shorter than those of modern Westerners, a third or more of the Hadza women outlive their fertility, going through menopause and in many cases becoming grandmothers.[26]

The Hadza and a handful of other tribes such as the Aché and the !Kung have been studied by researchers investigating evolutionary aspects of longevity such as "the grandmother hypothesis," the idea I touched on earlier — that women who live beyond their reproductive years and help care for the next generation are contributing to the survival of the species (or at least to continuing their branch of the family). If an older woman can help her children at the nest to the extent that there will be yet more offspring, or more who survive to adulthood to reproduce themselves, there will be more descendants to pass along the (fingers crossed) high-fitness genes that are a part of her genetic legacy. Grandmother is there to care for younger children, while a mother can focus on maintaining a new pregnancy and then nourishing the newborn.

The real work of grandmothering in ancestral times likely involved feeding the children who were weaned but not old enough to hunt and gather for themselves. The experiences of the modern-day Hadza, researchers believe, reflect those ancestral circumstances. The Hadza rely on tubers for nourishment—tough root vegetables that are buried deep in the dry, rocky ground. While a child can forage for berries and soft fruits, only a bigger, stronger person wielding a tool can extract the tubers from the hard earth, and that's a task for a grandmother with time on her hands (and a sharp digging stick in them, as well). While Hadza men and boys bring home meat from the hunt and hack honeycomb, another prized food source, from tree trunks, female elders are known to procure tubers to subsidize the food supply for a daughter's family.

Menopause provided a way for ancestral females to remain productive—beyond the reproductive years. You could say we evolved to live longer so that we could care for our grandchildren and ensure our genetic legacy through subsequent generations. Even with a few hot flashes thrown in, we're better off than Charlotte.

Advanced Eggonomics: The Midterm and the Final

A woman is not considered to be in menopause until her menstrual periods have ceased entirely for one year. (It really should be called meno-stop.) No period for four months in a row (and you're not pregnant)? If menstruation resumes, that's not menopause, yet. Nothing for nine months, and then a funky two-day period? Closer, but still not menopause.

The premenopausal state of perimenopause features symptoms that bookend those of PMS, in that they involve physical discomfort (hot flashes, vaginal dryness, trouble sleeping) and mood swings (along with low libido). As with all things hormonal, horror stories abound. *You will sweat through your bedsheets every night as you toss and turn...you are literally drying up "down there"...you are going to gain belly fat and look four months pregnant...you will grow a mustache and become a "handsome" woman....* If you read the life-as-you-know-it-is-over articles, a part of you may wonder if you shouldn't just join the ranks of the mayfly.

Yes, as you approach and eventually enter menopause your body will change — and your brain may, too. But much of what women experience as they transition to a new hormonal phase is simply part of human aging, gender aside, and not solely due to the disappearing act of the particular hormones that fertile women have in abundance. Women of all ages are done a disservice when the hype crowds out the facts, and when a lack of research on women's health contributes to a lack of information. (The cardiac risk associated with hormone replacement therapy — and with some formulations of the birth control pill — is real, but that "hormone hack" is a topic for the next chapter.)

Still, menopause is a reminder of how powerful our sex hormones are, and not just in women. There are two times in human life when it's difficult to distinguish between males and females, albeit from a distance (and with clothes on) — when children are small and when adults are elderly. Consider how hard it can be to tell the difference between a young girl and boy if they have the same haircut and are dressed similarly. (My mother cut my hair in a pixie cut when I was six, and I already had a "boy" name — so I can surely attest to this.) Fast-forward eighty years, after estrogen and testosterone have declined. Facial features may look less feminine or masculine, as the case may be, and — as in prepubescence — voice pitch may be indistinguishable. Childhood and old age have something in common: a dearth of sex hormones.

Menopause grants women longevity beyond the reproductive years, but I can't tell you with certainty why (or if) we evolved to have some of its accompanying symptoms, such as hot flashes or insomnia — unless we reframe these side effects as eggonomic trade-offs (I'm skeptical . . .). A number of studies show that women at high fertility prefer more masculine faces than menopausal women do, but menopausal women are more drawn to a wider range of "cute" male and female baby faces[27] than much younger (presumably fertile) women. In one study on "cuteness discrimination," as researchers termed it, women between the ages of fifty-three and sixty were more accepting of baby faces that were not traditionally adorable — by adorable, think big eyes, chubby cheeks, and all-around cherubic — compared to women in the nineteen to twenty-six age range (none of the women were mothers, so this looks like an effect of age).

So, in this eggonomics case, a grandmother may lose some estrogen-related benefits to her physical self, but psychologically, she gains an ability to be more accepting of other babies; here, she potentially embraces a broader latitude of offspring who are not her own. This will make her a better and more devoted helper at the nest if she's caring for extended kin (or even nonrelated offspring, whose families might help at the nests of her kin). A biological mother has an exclusive bond with her infant through sheer proximity as she protects and feeds her baby, but a menopausal grandmother won't possess the same intimate bonding mechanism, particularly with regard to lactation. Perhaps menopause offers an evolutionary assist here, by providing a grandmother with rose-colored glasses who will view all babies — even the ones who look like E.T. — as cute, and will help them survive and thrive. "A face only a mother could love"? It's more like a face every grandmother will love.

We experience a lifetime of estrous hormones, from the first stirrings of puberty through pregnancy and motherhood, should we take that path, and into our later years. One unfortunate bit of common ground across these three arcs is that we still tend to focus on the physical aspects of hormonal phases we have reason to fear or dread — messy menstruation, the drama of pregnancy (or lack thereof), menopause and its reminders that we've lost our fertility, the very thing that makes us female.

But it's a mistake to view normal, healthy hormonal phases as riddled with problems that need to be solved or symptoms that must be subdued. Certainly, each phase has its challenges (and we're not alone; men hit rough patches too), but so, too, does each phase hold deep pleasures made possible by the ebb and flow of female hormones. In fact, as you'll see, when we interfere too much with our natural cycles, we deny our hormonal intelligence, rather than allowing it to guide us through our lives as women.

8

Hormonal Intelligence

WHETHER SHE DOES SO unintentionally or by design, at some point in her life a woman will disrupt her own natural hormonal cycle, including estrus. Pregnancy, for instance, isn't the only cause of a missed period. Menstruation can be delayed, interrupted, or halted entirely by conditions such as extreme or sudden physical and emotional stress, drastic weight loss (or weight gain), illness, toxins in the environment, or breastfeeding. Most notably, the use of hormonal contraceptives or hormone replacement therapy will also alter the cycle, through changing the balance of key sex hormones and adding synthetic versions that mimic the body's naturally occurring ones.

It follows that if we engage in such "hormone hacking," involuntarily or on purpose, we potentially short-circuit our inherent hormonal intelligence, and in particular our ability to make strategic decisions regarding sexual behavior and mate selection. If a woman doesn't experience estrus and the evolutionary benefits it confers, will her hormonal intelligence still prevail, or will it be compromised? After all, messing with Mother Nature can have consequences.

It turns out that however powerful hormones may be — whether they are produced in the human body or in a pharmaceutical lab — they have their limitations in one regard: In searching for the existence of human estrus and confirming that it is real, we've also discovered that women evolved so that they would *not* be under strict hormonal control, so that they'd have free will, so that they'd be able to make strategic choices that would benefit their individual lives, if not choices that would perpetuate their genes.

Even if she chooses to break the cycle, or it is broken for her, every female still has hormonal intelligence across a lifetime. How she chooses to wield it is up to her.

Mother of All Hormone Hacks: The Pill

There is one form of estrous disruption that nearly 80 percent of women choose to engage in: hormonal birth control. Overwhelmingly, it is ingested in pill form, but the hormones involved — synthetic varieties of estrogen and progesterone (progestin) — can also be released internally through an implant or a ring, absorbed through a topical patch, or received via injection. When used correctly, the estrogen-progestin combination birth control pill and its cousins (e.g., "the ring") are exceedingly effective at preventing pregnancy. They also prevent or significantly reduce hormone cycles associated with estrus.

In many of the studies highlighted in previous chapters, as well as research conducted in my own lab, women who were using hormonal birth control were either not selected to participate or their results were analyzed separately, because we have good reason to believe that hormonal contraception can influence a woman's mate preferences. Many of the conclusions we've reached over the years about female estrous behaviors (such as an uptick in "roaming" behaviors, choosing more revealing clothing, or being more competitive), as well as about how women are perceived by others (she looks attractive, she smells good), are based on how a woman thinks and acts "at high fertility" — a phrase you've seen in these pages over and over again.

But hormonal birth control is designed precisely to suppress fertility and therefore stop pregnancy, meaning that there are *never* levels of "high fertility." In other words, when a woman chooses hormonal birth control, she is also choosing to forgo ovulatory cycle shifts. In the case of estrogen-progestin forms, the feedback loop that produces typical hormone shifts that cause estrus is effectively erased, and in that sense she is no longer "hormonal." In the pages that follow, when I refer to estrus-halting effects of hormonal birth control, *I am excluding the progestin-only minipill,* as well as other progestin-only forms; up to 40 percent of progestin-only users may continue to ovulate, an issue I'll address in a separate discussion. It is incorrect to say "the pill" stops estrus, as it depends on *which* pill.

Because hormonal cycles are essentially obliterated by estrogen-progestin contraceptives, some researchers have gone as far as suggesting that using these forms of birth control can impact a woman's attractiveness, cause her to make a genetically flawed mate choice (and therefore endanger future offspring), and trigger relationship problems, particularly with long-term partners. This is controversial, in part because it puts women in a damned-if-you-do, damned-if-you-don't fix with regard to using convenient and effective birth control. There is also the implication here that women are under strict hormonal control — even when they forfeit being hormonal! As you will see, it turns out that the story is not quite so simple (and some claims along these lines could be flat-out wrong).

How Not to Get Pregnant

Hormonal contraception can be divided into two main categories:

1. Contraception that contains a *combination of estrogen and progestin:*
    - A "combination pill": taken daily, twenty-one days of hormones, followed by seven days of a "blank" place-holder pill (which allow the body to briefly resume the cycle through menstruation[1])
    - A transdermal patch: worn for twenty-one days, then seven days off (menstruation) before replacing with a new patch
    - A vaginal ring: inserted and worn for twenty-one days, then seven days off (menstruation) before replacing with a new ring
2. Contraception that contains *progestin only:*
    - A minipill, so-called not because it is microscopic in size, but because it contains only one hormone; also, the dose of progestin in the minipill is considerably lower than that in the combo pill: twenty-one days of daily pills, seven days with no pill (menstruation)
    - An intrauterine device (IUD): inserted and worn for anywhere from three to five years (the hormonal IUD is not to be confused with the nonhormonal copper IUD)
    - An injection: effective for up to three months
    - A matchstick-size implant: inserted in the upper arm and worn for up to four years

Though my discussion will focus on "the pill," note that the patch, IUD, injection, and implant prevent pregnancy through the same mechanisms as oral contraceptives.

Estrogen-progestin contraceptives work by stopping ovulation. The ovaries stop releasing eggs for fertilization, and the uterine lining becomes thinner, making it difficult for any "rogue" egg to take hold. And just in case an egg does manage to make it out and travel down the fallopian tubes en route to the uterus, progestin causes a thickening of cervical mucus that stops the sperm from entering the egg. The idea behind this double whammy is simple: Ideally, estrogen makes sure nothing gets out while progestin makes sure nothing gets in.

With progestin-only birth control, ovulation may stop for some women, but it has been reported to occur up to 40 percent of the time. That's why the sperm-reducing Cervical Bermuda Triangle, courtesy of progestin, is so important if pregnancy is to be prevented.

In the absence of hormonal contraception, the consistency of cervical mucus changes across the ovulatory cycle. As peak fertility approaches, cervical mucus becomes thin and plentiful and provides biochemical nourishment for sperm, which can live for several days in the fallopian tubes awaiting an egg. To facilitate fertilization, the mucus becomes more slippery and stretchy (like egg whites, fertility experts like to say), allowing the sperm to travel easily through it to reach the egg. Cervical mucus can also filter out structurally abnormal sperm (like the bouncer at an exclusive nightclub — "You look good, you're in. Nope, you look weird and sick — stay out").

But progestin-only contraceptives such as the minipill can turn cervical mucus into a brick wall and quicksand, all at once. Progestin thickens cervical mucus, as discussed, stopping the sperm's journey. When sperm try to storm the barricades to reach the egg, they fail. Should any make a dent, they're trapped in the cervical mucus and are never heard from again. Progestin versus sperm? It's no contest — the boys are on a suicide mission. (Granted, even without contraception and on a good baby-making day most of the tens of millions of sperm released will die. But all it takes is one.)

For some women, the side effects of estrogen-progestin or progestin-only contraceptives are the deciding factor in choosing whether they'll take the combination pill or the minipill. Because the minipill contains a lower dose of progestin, it isn't as "strong" as the combo pill — nor is its impact on estrus, but some prefer it because it is not associated with blood clots, stroke, or heart disease, unlike the estrogen-containing combination pill. Progestin may also be a safer bet for smokers than the combo pill, and for anyone who is breastfeeding, it won't impact milk production. (Yes, breastfeeding suppresses ovulation and therefore is a form of birth control — but not always. "Irish twins" aren't always Irish, and they are never twins. For more on breastfeeding and how it impacts the hormonal cycle, see "Tiny Hungry Hackers: Childbirth and Breastfeeding," later in this chapter.)

There are numerous brands and formulations of oral contraceptives, as well as generic versions, and like any medication, all have possible side effects. If a woman's medical history and current state of health are not an issue, and if she has no preference, her healthcare provider may simply prescribe whatever formulation he or she is most accustomed to (or perhaps had been sold most persuasively by the drug distributor). If taken as directed, both pills achieve the same contraceptive goal — but that's where their similarities end, because the combination pill obliterates the "high fertility" phase and its related behaviors.

The estrous switch gets flipped to off through estrogen-progestin hormonal contraception in large part because gonadotropin-releasing hormone (GnRH), the ovulation Stage

Manager discussed in Chapter 3, is no longer secreted from the brain. GnRH triggers the initial and crucial rise of follicle-stimulating hormone (FH) and luteinizing hormone (LH), which are responsible for the release and maintenance of a healthy egg suitable for fertilization. Without GnRH, FH and LH are still present in the system, but their levels remain flat across the cycle since ovulation is not occurring. Similarly, the naturally cycling ups and downs of estrogen and progesterone are replaced by a flat level of estrogen and a plateau-like level of progestin. The following diagram shows a normal cycle, above, and a cycle transformed by the combo pill, below.

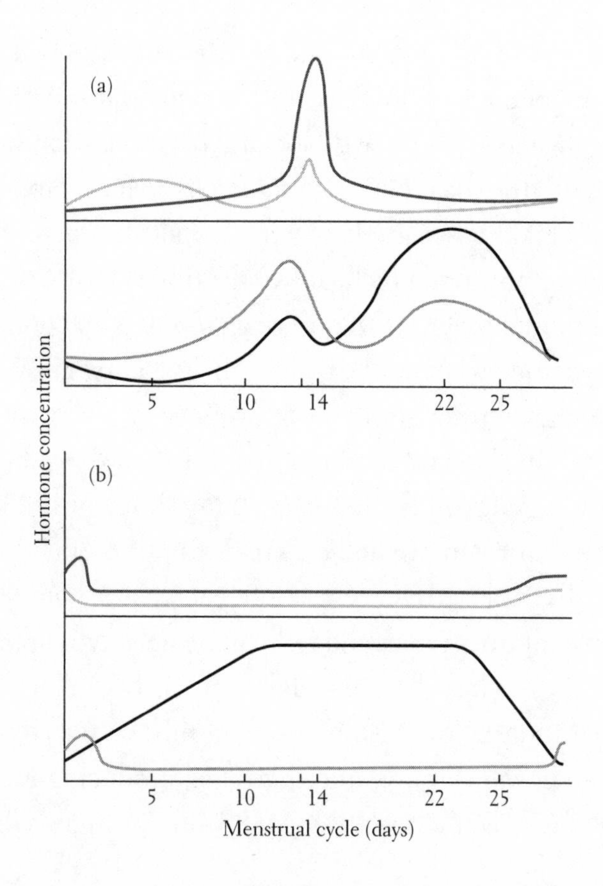

The combination pill disrupts ovulation and estrous behaviors. It is unlikely that women consider the differences in oral contraceptives when they're filling their prescriptions, unless they've made a conscious decision based on side effects or health history. It is also unlikely that women are asking this question: To what extent does the combination pill (or any other estrogen-progestin contraceptive) blunt hormonal intelligence and impact the strategic behaviors that can guide a woman's decision making — and her sexual and social fate? Let's explore that question and some possible answers.

Mate Shopping on the Pill: Looking for Mr. Just Like Me

Given that estrous desires can nudge females to seek out potential mates with good genes, it is easy to conclude that women call off that search when taking estrus-blocking hormonal contraceptives. At least, that's the idea that some researchers[2] have proposed, and they point to it as an example of why the pill may be problematic when it comes to producing healthy offspring. This theory is based on some findings that women on estrogen-progestin contraceptives are drawn to men who may be genetically incompatible — because they are too genetically similar.

Earlier, I explained the role of major histocompatibility complex (MHC) genes (see "Kiss-Kiss, Bang-Bang," in Chapter 6), and why parents with *dissimilar* MHC might produce healthier offspring; dissimilar MHC genes from each parent strengthen the offspring's immune system and reduce the negative consequences of inbreeding, such as disease and deformity. Take a look at Goya's famous paintings of the Spanish royal family, with its tendency to consolidate power through intermarriage, producing inbred couples, and you'll see the results of MHC-similar unions dressed in regal silks and satins. But, palace intrigue aside, all this begs the question, Why would human females be programmed to seek out males with dangerously similar genes to father their offspring (and doom their health)?

One theory is based on the idea that when the pill stops ovulation, it sends the body into a state resembling early pregnancy, particularly during the latter half of the cycle, when progesterone is especially high. Some scientists believe that pregnant women, who have extremely elevated levels of progesterone and who are at their most vulnerable, show a preference for close contact with their own kin — with whom they share genetic material — rather than strangers. If a potential mate materializes for a woman on the pill, in her not-pregnant-but-my-body-thinks-I-*might*-be state, she may be drawn to him if she detects a hint of MHC similarity.[3]

There are a few problems with this explanation, however. While some formulations of the pill may cause the body to mimic certain aspects of pregnancy (notably the lack of ovulation), the amount of progesterone the body produces during pregnancy is much higher than the amount of synthetic progesterone delivered through hormonal contraception. The idea that the pill creates a faux-pregnancy state that results in genetically similar couples is an oversimplification.

When a handful of studies from about a decade ago first suggested that a woman's mate-choice radar went haywire due to hormonal contraceptives, the findings made the headlines, and the tone was one of near panic, with alarmist messages such as "How the Pill Could Ruin Your Life"![4] Aside from the problem of women choosing MHC-similar partners, went the popular narrative, they were also choosing less masculine men, because they were never reaching high fertility. They weren't drawn to more "manly" men with symmetrical features and deeper voices; they were drawn to some of the classic ancestral indicators of high-fitness genes.

If a woman uses the pill, whither the Sexy Cad? Was the pill (invented by men) the revenge of the nerds?

Relationship Quality: Mr. Right and Ms. Whatever

The studies on mate shopping and the impact of hormonal birth control twisted the knife with another romance-busting suggestion — that women in committed relationships who met their partners *before* they began using hormonal contraceptives would take a dimmer view of their mates once they began doing so. Estrus would cease, and so would a woman's sexual attraction to her boyfriend or (too late!) her husband. The fires of passion didn't simply go out when a woman took the pill — hormonal contraception prevented them from igniting. *Oh, it's you again.*

It could cut the other way as well, scientists predicted. For those women who met their partners while they were on estrus-blocking contraception and then stopped using it in the midst of their now-established relationship, the scales could fall from their eyes and they could truly see their male partners for how ho-hum they really were. *Why was I sexually attracted to him?* Now that she had ceased taking hormonal birth control, the thinking went, a woman might experience estrous urges and be attracted to a different type of male — a manly, symmetrical, Sexy Cad brimming with good (much better than her current partner's) genes... or at least she'd feel that way on a handful of high-fertility days, and perhaps some women would even act on those attractions and seek other partners, at least within the fertile window.

There seemed to be no end to this newly discovered minefield of theorizing about the effects of the pill, and as long as these ideas were piling up, here was one more prediction: The pill could make a woman unattractive and fat (no more of those calorie-burning, wheel-running days in the middle of the cycle). Extrapolating from the findings on estrous behaviors, scientists suggested a woman without a high-fertility phase would no longer be inclined to engage in "ornamentation" and energy-expending, calorie-burning roaming behaviors. (I imagine that in a lab somewhere, a researcher added an estrogen-progestin cocktail to an estrous rat's water supply, just to see if she'd stop running on her wheel and put on a few ounces.) Yes, the pill could cause a woman — now in a quagmire of a relationship with Mr. He's Okay I Guess — to camp out on the couch in an oversized stained T-shirt with a bag of cookies. (And she probably wouldn't feel terribly competitive, either. So much for job performance, not to mention her tennis game. . . .)

The pill was being recast as potentially poisonous, but I questioned these sweeping generalizations on a woman's mate preferences as well as the conclusions about poor mate choice or damaged relationships — because the evidence was just beginning to emerge.

I was convinced of just how incomplete the evidence was by my PhD student Christina Larson, who forcefully criticized the oversimplified messages about the pill, especially given just how liberating reliable contraception can be for women. Her dissertation[5] was an impassioned but careful review of every study pertinent to the question of whether women's mate choices differed depending on whether they used hormonal contraceptives or not. She showed that the evidence from those studies was quite weak, results were mixed, and some of the research, which dates back at least twenty years, did not use the more rigorous methods that were developed later (and that yielded different results). For instance, not all the experimenters testing women's preferences at high fertility excluded users of hormonal contraceptives, or differentiated between women who were and were not using them. If researchers did separate women using hormonal contraceptives from those who did not, they collapsed different pill mechanisms (e.g., the combo pill, which halts estrus, and the minipill, which only halts sperm and not always ovulation). Moreover, Christina pointed out a fundamental flaw in the logic that we'll choose the wrong long-term mates when using hormonal contraception — nearly all of the past research shows that shifts in women's preferences across the cycle are specifically for *short-term* mates, or mates women deemed attractive in the immediate moment, not whom they picked to stick around long-term. Given this pattern, using the pill should not lead long-term mate choices to differ very much.

Contraception, Contradiction, and Some Conclusions

Christina and I wanted to test the idea that the pill (and other forms of hormonal birth control) might lead women to select genetically incompatible men.[6] So we set up a study using a large sample of couples in committed relationships, including married partners. One would expect that relationships that started when women were already on the pill (when the estrous switch was flipped off) would feature couples who were more MHC-similar, compared to relationships that started when women weren't on the pill. Unexpectedly, we found a trend in the opposite direction.[7] Partners in couples who initiated their relationships when the woman was already using the pill were more MHC-*dis*similar from one another than partners in couples where the woman began using the pill after the relationship was established. This threw a wrench into "the pill makes women choose the wrong men" narrative.

Around the same time that we were conducting this research, a methodologically rigorous study appeared that showed that women who started their relationships while on the pill (all were on the combination pill, with only one exception) experienced a decrease in relationship satisfaction when they discontinued it, but only if their male partner was relatively low in facial attractiveness (as rated by independent coders).[8]

In sum, although the pill might not lead women to choose MHC-similar partners, whether a woman starts her relationship on the pill and then resumes regular cycling could have an impact on how she views her partner's facial attractiveness.

What about shifts in women's attractions across the cycle? If women using the pill do not experience the estrous desires that draw them to the alpha-male types a few days out of the month, what are the consequences?

In a second study from Larson's dissertation, she examined just this. She followed women and their partners over the course of a month, comparing couples in which the women were on the combination estrogen-progestin pill with those using no hormonal contraception. As in our prior research, she found that naturally cycling women who evaluated their male partners as relatively low in sexual attractiveness experienced an increase in attraction to *other men* at high fertility. Women who were taking the pill showed no upward shift in "extra-pair" attraction. This raises the question of whether hormonal contraceptives could in some ways be relationship protectors. The consequences of the pill could be there, but the story is not quite as simple as "it mucks up mate choice."

New research from Norway[9] creates another wrinkle in the story, and it offers women opportunities for nuanced hormone hacking. Researchers concluded that different formulations of hormonal contraception can have quite different impacts on our relationships, because contraceptives with higher doses of progestins are more extended-sexuality-like, whereas those with higher doses of estrogen are more estrus-like. And, indeed, women taking higher doses of progestins, and who feel strong loyalty and commitment to their steady partner (perhaps those Good Dad guys), have more sex with them throughout the cycle, following an extended-sexuality pattern. Women who take pills with higher estrogen concentrations show a pattern that runs in reverse — the more loyal and faithful they feel to their partners, the *less* sex they have with them (perhaps because they are on the lookout for Mr. Sexy instead of Mr. Stable).

My own view is that the hormonally intelligent thing for a woman to do is to use the best means to control her fertility as she wishes to do it. Perhaps this means a longer conversation with the ob-gyn about the hormone concentrations in each of her contraceptive options. Perhaps it means she tries different formulations to see which make her feel best and how they affect how she feels about her partner. Or, perhaps she chooses not to alter her hormones (using a nonhormonal IUD, for example) and to ride the waves of estrus.

G-String Theory

Those who subscribe to the idea that women on the pill are less attractive to others at midcycle than their naturally cycling sisters point to an oft-cited 2007 study on eighteen lap dancers—some of whom were on the pill—and the amount of tips they earned as they approached ovulation and high fertility.[10] Since lap dancing involves intimate but semiclothed contact between the genital areas of the dancer and her male patron, it's difficult to think of a better (legal) way to test the real-time attractiveness of a woman at midcycle, on the pill and off.

As it turns out, over the course of sixty days (and two cycles), the naturally cycling dancers outearned their contraceptive-using colleagues—earning about eighty dollars more per five-hour shift in tips when they were at their most fertile and when ovulatory cues would have been detectable. Furthermore, non–pill users experienced a rise and fall in hourly average earnings depending on where they were in their cycles—seventy dollars in estrus; fifty dollars in the luteal phase; thirty-five dollars during menstruation. Both groups of dancers earned less when they were on their periods, but the dancers on the pill saw no spike in their earnings during ovulation. For those who wanted to cherry-pick a big, juicy conclusion from this study, here it was: Not only did the pill make women unattractive; it was also making them poor.

Hardly. Although I am a big fan of research that follows women over time to get better assessment of change across the cycle (and does so "in the wild"), the study was tiny and we don't know important things about the women in it—like which hormone formulation they were using. It's a good bet that most or all were on the combo pill, since it is used much more often. But also consider the economic consequences of women not using reliable birth control, frame this in terms of long-term costs and benefits (rather than short-term attractiveness), and look closely at the bottom line. The socioeconomic benefits of effective hormonal birth control far outweigh the high costs of not being able to control your career and earn a good income (unless you're a lap dancer).

Tiny Hungry Hackers: Childbirth and Breastfeeding

Doctors generally tell new mothers at their six-week postpartum checkup to begin using birth control whether or not they are breastfeeding. This news is usually delivered by an ob-gyn with a straight face, to a woman who may have a *Sex? Are you kidding?* look on hers. There is a good chance the mother is carrying an overstuffed purse crammed with new-baby paraphernalia, plus various shapes and sizes of absorbent pads for her own use, as she's still springing leaks here and there. Though she listens to the doctor's advice, she's thinking the age-old thought — that recent childbirth and having a newborn are their own form of contraception, and sex is not on the table, literally or figuratively, at least until the baby sleeps for more than three hours at a stretch.

Like hormonal contraception, pregnancy, childbirth, and breastfeeding disrupt the ovulation cycle dramatically and with an obvious purpose — to ensure reproduction and the nurturing of offspring. It generally takes several full cycles postpartum before hormones reset to pre-pregnancy levels and the entire ovulatory process, including menstruation and the ability to become pregnant again, resumes. (Contrast that fertility time frame with fast-reacting progestin-only contraception; with the progestin-only implant and IUD, for instance, pregnancy is possible within weeks or even days of its removal.)

Nature's Birth Control

Breastfeeding can be a form of birth control because it temporarily halts ovulation. Prolactin, the milk-making hormone, helps to suppress estrogen and other hormones that contribute to fertility. The more a mother nurses, the higher her levels of prolactin. (This is why breastfeeding experts tell mothers that to make more milk, breastfeed "on demand" or "on cue" whenever the baby is hungry, and this will keep prolactin levels high and the milk supply flowing.)

In order to use the "lactational amenorrhea method" (known as LAM) as effective birth control, however, a mother must be exclusively breastfeeding her baby on demand — "exclusively" meaning no food other than breast milk, and no bottles of supplemental formula, or even breast milk. Pumping breast milk rather than nursing is thought by some experts to be less effective in keeping prolactin levels high, and giving bottles can lead to the infamous "nipple confusion" (so no pacifiers, either). With LAM, a baby should be fed regularly and frequently for a mother to maintain adequate prolactin — ideally every couple of hours, including night feeding. Skipping a session and offering a bottle of expressed breast milk or formula from time to time may not seem like an oops-causing contraceptive slipup, but as soon as prolactin levels dip, estrogen revs up again in an effort to return the body to a natural ovulatory cycle. If that dip happens often enough, full fertility may return and breastfeeding as birth control would no longer be effective.

In short, using LAM to prevent pregnancy requires commitment and a flexible schedule. It probably worked well for ancestral women who were able to devote their full attention to caring for offspring. (They did not, after all, have to go back to work after maternity leave.) Anthropologists estimate that human children were not fully weaned until nearly three years of age,[11] though by then they were eating other foods and the fertility-suppressing effects of breastfeeding were probably no longer in full force. Still, given that ancestral mothers had to carry their babies everywhere until they learned to walk on their own, it makes sense that we evolved to give mom a break between pregnancies with this bit of natural birth control.

Eventually, even among mothers who breastfeed beyond the American Academy of Pediatrics' six-month recommendation, menstruation will return. (The AAP recommends six months of exclusive breastfeeding, and at least another six months of additional breastfeeding, till the baby reaches the age of twelve months, along with solid foods and, if used, supplemental formula.) In fact, ovulation can resume before periods do, because fertility returns in stages after childbirth as natural hormonal levels come back into balance. For instance, there may be enough follicle-stimulating and luteinizing hormone to trigger maturation and release of an egg, but no period will follow if estrogen and progesterone levels are still low; in that case, the uterine lining is "incompetent," as some doctors say, meaning that it can't support and nourish an egg for implantation. (But I have to ask: Would they call a man's testicles incompetent if he were recovering from a vasectomy?) Conversely, it is possible to have a period without ovulation, if the uterine lining is once again competent, but there is no egg.

In the same way that fertility returns in stages, so, too, does estrus, as a woman gradually shifts from mommy mind back to mating mind, and the natural cycle resumes.

The Other Breastfeeding Hormone

Along with prolactin, the hormone oxytocin also rises with breastfeeding. Oxytocin is responsible for the let-down reflex nursing mothers experience, as it helps to release milk into the ducts. It also causes uterine muscles to contract after childbirth; when a new mother nurses her newborn, the oxytocin produced helps shrink the postpartum uterus to its pre-pregnancy size.

Oxytocin, popularly described as "the cuddling hormone," has another important function postpartum. It is a key social-bonding hormone — it heightens a mother's desire to defend and care for her infant, and as such it is a major contributor to the Mama-Bear Effect (see "The Mama-Bear Effect," in Chapter 7). Certainly, non-breastfeeding mothers can be ferociously protective too. And oxytocin is released not only during lactation but also during childbirth itself and in interactions with a newborn. But breastfeeding seems to reinforce the Mama-Bear Effect, perhaps conferring an extended level of protection for offspring, as oxytocin helps mothers remain aware of threats from attack, accident, and disease beyond the immediate aftermath of birth.

Besides its role in milk production and maternal bonding, oxytocin could be part of a postpartum cocktail (perhaps along with prolactin) that functions as a kind of natural antidepressant, with breastfeeding mothers showing lower levels of postpartum depression (PPD) compared to non-nursing mothers. The Centers for Disease Control and Prevention estimate that about one in nine new mothers report symptoms of depression.[12] Unlike transitory feelings of depression, PPD does not "go away" on its own; it usually requires treatment. (It is normal for new mothers to feel a wide range of emotions after the birth of a child, including some depression — but those temporary "baby blues" are not the same as PPD.)

My former postdoctoral student Jennifer Hahn-Holbrook and I explored the connection between PPD and breastfeeding, examining existing data that showed lower levels of PPD among nursing mothers.[13] Though some research suggests that breastfeeding may mitigate depression, we believe the situation is more complicated. We wonder if perhaps PPD isn't another "disease of modern civilization," much like diabetes or heart disease. We believe it's likely that PPD didn't exist (or was rare) among ancestral women, as they didn't have the same kinds of stressors in new motherhood as we do.

For starters, ancestral women probably didn't go it alone. They lived in family and social groups with extended kin, meaning that help was never far away, unlike modern families, which can be spread across many time zones. Being alone with a vulnerable infant in your care — even with a partner — can feel overwhelming at times, especially without a support network.

In our review of the research, other stressors faced by modern moms that caught our attention include: diets lower in inflammation-fighting omega-3 fats, which have been shown to help in alleviating depression (pregnant women are discouraged from consuming omega-3-rich fish, due to concerns over mercury); low levels of sunlight, which produces vitamin D, another anti-inflammatory nutrient; and less physical activity, which can have a negative impact on emotional well-being. Finally, some women choose not to breastfeed or they may be unable to. Working moms who nurse during maternity leave are likely to start weaning their babies or shift to pumping breast milk within the first six months. We hesitate to say that breastfeeding necessarily alleviates PPD or is the best course for all moms — it can itself become a debilitating stressor (lack of time, lack of sleep, lack of support in the workplace or at home). Moreover, there is no conclusive evidence that a mother becomes depressed when she weans her baby.

Postpartum depression, as well as the normal mood swings new mothers experience, is too often another excuse to label women with the h-word. *Don't call her this week — she just had the baby and she's so hormonal! She'll cry or get angry at the drop of a hat.* A new mother doesn't need to be judged for whether or not she breastfeeds, and how "successful" she may be if she makes the effort to do so. Instead, she needs unconditional support, especially in the absence of extended family and the close kinships our female ancestors benefited from. Breastfeeding certainly might help, but it's not the only answer.

In my personal experience, breastfeeding helped me transition to motherhood. I had my doubts about trying to nurse twins. I was a little put off (and scared) by some of the dogmatic guidelines from hard-core lactation advocates. And would I really have to produce a gallon a day? (Yep.) But I became a true believer once I experienced how easy it was for me to breastfeed to calm a crying infant (or two: *thank you, special pillow*), how close I felt to them, and the gentle buzz of warmth and calm I felt while having them at my breast. I also felt a kinship with the long line of female ancestors who had done the same. (And, bonus: With a 50 percent increase in my need for calories, food had never tasted quite so good.) I was privileged to have the support at work and at home I needed in order to breastfeed two infants. It worked for me, but certainly breastfeeding isn't the only "right" way — I know it's not for everyone — as many loving mothers fully nourish their babies without breastfeeding.

Breastfeeding, for me, actually became a way to adjust to the demands of motherhood. But there are other ways, of course. Fortunately, just snuggling with your little one with lots of skin-to-skin contact, leaning on your experienced mom and dad friends as well as your partner (if you have one), and taking care of your own physical and emotional needs — a little sunlight, physical activity, and naps work wonders — are all effective ways to refuel your body and replenish your spirit as you take on the life-changing role of parent.

And if you're not a mother, then you can easily become part of the social support network moms need: Hormonal intelligence isn't just about how the brain and body are transformed by childbirth and breastfeeding. It is also about being more supportive and compassionate to any woman — sister, stranger, colleague, or employee — when she makes the transition to motherhood.

The Hormonal Intelligence of a Real Shrew

The common tree shrew, found in Southeast Asia, is a small but smart mother. She always gives birth to twins. But adult tree shrews build two separate nests: one for their offspring and another for mom—and dad, as it turns out. Tree shrews are usually monogamous, so the male is there to help at the nest when the babies arrive. In fact, *he* builds the nest. After her twins are born, the mother retreats to her "no kids allowed" home but returns to nurse her babies. Still, she only pops in on them for a mere ten to fifteen minutes at a time, and only every two days (it's extremely nutritious milk). Then she skedaddles back across the branches to her own nest. Talk about carving out some "me time"! Once the twins are about three months old and weaned, they are allowed into the parents' nest to live until they each head out to find their own mates, who will presumably be primed for the same co-parenting arrangement.

Pills and Potions: Hormone Replacement Therapy

If you think that today's consumer-targeted advertisements for erectile dysfunction drugs seem cringe-worthy, then consider vintage print ads for Premarin, an early hormone replacement therapy drug approved by the FDA in 1942 for treating common menopause symptoms, still prescribed by doctors more than seventy-five years later.

"The Calm of Eventide..." says the quaint headline of an early ad, featuring an illustration of a contemplative woman shown in the cool, bluish shadows of early evening — eventide. She looks as if she has found inner peace (through a synthetic estrogen made with the urine of pregnant mares) despite "the change," which evidently makes women go utterly bonkers. Not this formerly manic lady. She's had her Premarin and she's relaxed.

A 1960s ad features a happy-looking couple aboard their sporty sailboat. "Husbands Like Premarin, Too," its headline says, as the master and commander himself grins at the camera, his wife gazing up at him adoringly. (She found the Calm of Eventide!) The copy says that a doctor who prescribes Premarin to his patient during menopause "usually makes her pleasant to live with once again." Because before she started HRT, she was such a nag.

The ad, probably directed at physicians since this was before prescription drugs were routinely advertised straight to consumers, goes on to say that it's hard enough for men "to take the stings and barbs of business life, then to come home to the turmoil" of a menopausal spouse. (After all, she is one crazy, hormonal bitch.)[14] But start her on Premarin and the missus "is a happy woman again, something for which husbands are grateful."

To address the more uncomfortable symptoms of menopause,

especially hot flashes, night sweats, and vaginal dryness, women for decades have been seeking out solutions, including HRT drugs such as Premarin and many other brands and formulations. (Like hormonal contraception, HRT can emphasize one hormone — in this case, estrogen — or offer a balance of estrogen- and progesterone-like compounds.) HRT is effective, but this particular hormone hack is not without controversy and, for some women, considerable health risks. The road to the Calm of Eventide has not been a smooth one.

Feminine Forever: Better Living through Chemistry

Like puberty, PMS, menstruation, and pregnancy, menopause is another hormonal phase that the medical community, pharmaceutical companies, and some wellness experts seem to frame primarily as a problem to be solved, one more messy lady-parts phase. The pioneering use of HRT was, broadly speaking, a positive turn of events in women's health, but initially it seems the drugs were routinely prescribed by male doctors almost as if "the change" was a disease in need of a treatment, and that psychological — not just physical — symptoms of "the menopause" were dire, including "psychic instability." (Here are other Premarin headlines from their decades-long print campaigns: "When women outlive their ovaries" . . . "Something is terribly wrong" . . . "Her family was bewildered.")

In 1966, a New York gynecologist named Robert A. Wilson published a bestselling book called *Feminine Forever*, portraying the loss of estrogen as an avoidable tragedy. Some choice quotes include "menopausal castration amounts to a mutilation of the whole body"; in discussing symptoms he refers to "the horror of this living decay"; a menopausal woman is "no longer a woman but a neuter."[15] The answer? Estrogen replacement with synthetic hormones, which he claims to have "dreamed up," a problematic boast since his book was published more than twenty years after FDA approval of Premarin. Wilson, it turns out, had financial stakes in the pharmaceutical companies that manufactured synthetic hormones for HRT, though this would not come to light for many years. Meanwhile, *Feminine Forever* worked its wiles on doctors and patients alike, as it contributed to the trend of medicating menopause. Prescribing HRT for women who had entered menopause was becoming a routine practice.

Less than a decade earlier, young women had received the first prescriptions for the birth control pill. Now their mothers had a pill of their own — sometimes called "the youth pill."[16] In both cases, for both generations, the damage to women's well-being from poorly understood pharmaceuticals — whether their symptoms were psychological or physical — was under way.

Bad Medicine

Even years after the sexist, Mad Men–era ads faded away, HRT and "the change" were discussed in urgent tones, tinged with fear and dread. Television ads in the 1980s and '90s featured concerned women speaking to their doctors or to one another and discussing the horrors of menopause beyond its uncomfortable and embarrassing symptoms, including its link to bone loss, heart problems, Alzheimer's disease, colon cancer, and even tooth loss and blindness. HRT, women were told, would put the brakes on this hellish passage. In fact, starting in 1992 the American College of Physicians officially recommended it as a preventive therapy, especially against coronary heart disease, osteoporosis, and dementia.[17]

But HRT, it turns out, was not a silver bullet. On the contrary, some women were destroying their health by using synthetic hormones. Beginning in the late 1990s the Heart and Estrogen/Progestin Replacement Study (HERS) and the Women's Health Initiative study revealed that women who had been in menopause for more than ten years and who were on HRT were *increasing* their risk of heart attack, stroke, and blood clots. In 2002, a broad clinical trial involving women taking HRT and a group taking a placebo was halted because those on hormone therapies had a *higher* risk of breast cancer. (And it wasn't just "risk." Lawsuits against drug makers would follow, brought by HRT patients diagnosed with heart disease and cancer.)

Initially, women were advised to stop estrogen therapy because its cardiovascular risks outweighed the benefits. To relieve symptoms such as hot flashes, some women turned to alternatives, including soy foods (for their phytoestrogens — weak, plant-based estrogens), herbal remedies such as black cohosh, and bioidentical hormones from special compounding pharmacies — considered more "natural" than synthetic hormones produced on a large scale. But those solutions were often flawed and ineffective as well, unregulated treatments that carried their own health risks. (Bioidentical hormones, for example, can vary widely in quality and effectiveness depending on the compounding pharmacy.)

Now doctors and researchers have confirmed that the timing and type of estrogen therapy matter. For instance, during the first six to ten years of menopause, taking estrogen can lower your cardiac risks. But after the ten-to-twelve-year mark, estrogen can increase your risk of heart attack or stroke.[18] Drug makers have also reformulated their products over the years in an attempt to lower risks. (As with hormonal contraception, pills remain the most popular form of HRT, but it may also be dispensed in topical patches, creams, sprays and gels, vaginal rings, and suppositories.) The bottom-line advice everyone seems to agree upon: When it comes to HRT, take the lowest effective dose for the shortest amount of time.

Perhaps if more research in the lab had focused on women's issues in the decades prior to the revelations of serious health risks, the rollout of HRT would have been considerably safer. It is possible to use HRT safely and successfully, but the lesson is clear: There is no one-size-fits-all way to manage menopause, because every woman has a different hormonal experience, across her lifetime. Framing menopause — or any other normal hormonal change — as an illness in need of a cure is the wrong approach, one that has not served us well.

The End of Estrus, Not the End of Hormonal Intelligence

Feminine Forever was a popular book in part because it tantalized readers with the prospect of sex — estrogen therapy would restore a woman's desirability and her desire. Wilson promised that "bodily changes typical of middle age can be reversed." And the sex would be nonstop, right up until the grave. "Now, almost any woman, regardless of age, can safely live a full sex life for her entire life," the front cover announced. It was all hot estrus, all the time! But if husbands were worried that their estrogen-flushed, horny housewives (cougars) would jump the refrigerator repairman, Wilson was there to assure them with his own definition of extended sexuality and pair-bonding: "an estrogen-rich woman capable of being physically and emotionally fulfilled by her husband... is least likely to go afield in search of casual encounters."[19]

This vision of a pliant and faithful woman, able to satisfy her man, seems to be consistent with the era — the 1960s, when women were beginning to raise their collective voice and men were constantly trying to shut them up. (Don't forget our other doctor-friend, whom we met back in Chapter 1 — Edgar Berman, who cautioned in the early 1970s that because women were subject to "raging hormonal influences" they could never achieve equality with men.)

This old take on menopause and HRT was based on the belief that a woman was under strict hormonal control; therefore, when sexy, feminine estrogen dried up, so did she, in mind and body. The medical establishment's answer was simply to make her hormonal once again, with synthetic compounds that would prove to be dangerous. At least in the short term, before her health was compromised, she was able to "live a full sex life" and make her husband happy.

Estrus begins with puberty and ends with menopause, but to say that a woman's sexual urges stop when her fertility cycle does is nonsense. Perhaps we are no longer nudged by estrous behaviors when we reach menopause, but our sexuality is no longer driven by an ancient need to find a good-genes mate and reproduce; it is driven by arousal, desire, and closeness to our partner.

At this stage, then, we are entirely independent and in control, and our hormonal intelligence ultimately evolves into something new: wisdom.

(Un)Natural Woman: Resetting the Biological Clock

Though I've emphasized that the "average" age for girls and women to hit certain hormonal milestones is variable, there is a range that is outside the norm. Being a modern female and dealing with having a baby at age eighteen or forty-five is one thing, but coping with growing pubic hair and breast buds at age eight, or periods at age nine, is another. Early puberty, more prevalent among girls than boys, is increasingly common, and it seems to be related to a variety of twenty-first-century factors.

For many decades doctors considered eleven years of age "average" for the onset of puberty in girls — the beginning of physical developments, including pubic and armpit hair, breast growth, and menstruation, that culminate in the ability to become pregnant. But in 1997, a study published in the journal *Pediatrics*[20] revealed that among a sampling of seventeen thousand girls in the United States, the average age for breast bud growth was just shy of age ten for white girls, and slightly younger than age nine for black girls. The lead author of that landmark study, Marcia Herman-Giddens, had been working as a physician's associate in pediatrics and had noticed very early physical developments, including breast development, in eight- and nine-year-old patients. (Herman-Giddens is now a professor of maternal and child health.)[21]

At first, it seemed no one wanted to consider the possibility that American girls were entering puberty so early, including pediatric endocrinologists, and Herman-Giddens's research was ignored or questioned. But parents of girls who were experiencing early signs of puberty embraced the news-making findings, which confirmed their own observations of their daughters' somewhat alarming physical developments. In 2010, another study published in *Pediatrics* revealed even more stunning data, and the scientific community agreed that early puberty was real: Researchers noted early breast growth in more than 10 percent of white girls, 15 percent of Hispanic girls, and 23 percent of black girls — by age seven. "The proportion of girls who had breast development at ages seven and eight years, particularly among white girls, is greater than that reported from studies of girls who were born ten to thirty years earlier," the authors of the study concluded.[22]

There is no longer any debate as to whether or not early puberty is happening, but there are many questions as to why. Theories abound and include potential toxins found in our food chain and chemical compounds in everyday products. Some studies have linked family stress, including maternal depression and the presence of a stepfather, to early puberty. Perhaps in ancestral times, early maturation as a response to a hard childhood made sense; the sooner you could become independent, the better (and better not to become an evolutionary dead end if your future was uncertain). But we live in a modern world, and the cost of leaving home at too young an age surely outweighs the benefits.

The hunt is on for external factors that are causing earlier changes to the young brain, where puberty, and a lifetime of hormonal cycles, originates.

Environmental Factors

If you have a daughter, a little sister, or another young girl in your circle of friends and family, you have probably heard about toxic chemicals that can contribute to early puberty, including bisphenol A (BPA), used in plastics manufacturing from plastic food wrap to the linings of metal food cans to cash register receipts. BPA is molecularly similar to estrogen, and some scientists think that it can produce similar effects in the body, such as breast growth in young girls.

When consumers started to learn of the health risks that BPA posed to children and adults (studies show possible links between BPA and everything from cancer to dementia), big companies began removing it from their products — it is no longer found in baby bottles and sippy cups — and offering substitutes. (The "BPA-free" label may now be as ubiquitous as BPA itself once was.) But there is evidence that the substitutes are not necessarily safer.

Zebra fish exposed to low levels of an alternative bisphenol, known as BPS, showed the same disruptions in their reproductive patterns as they did with BPA exposure, including accelerated development of embryos and faster hatching patterns. Like BPA, BPS affects the neuroendocrine system, including the secretion of GnRH, the hormone that helps to kick off puberty and, ultimately, fertility. In other words, chemicals such as BPA and BPS speed up key developments of the reproductive system — and they may have the same impact on humans, including young girls.[23]

BPA received tremendous attention, but other estrogen-mimicking compounds are also present in some detergents, pesticides, flame retardants, phthalates (a category of chemical "plasticizer" found in everything from personal-care items to flooring), and other products. While some compounds may have little to no effect on early puberty, others appear to be quite powerful. No ethical study would ever subject a child (or an adult) to potentially toxic chemical exposure, but the evidence exists. In the early 1970s, cattle in Michigan ingested grain that had been accidentally contaminated with a flame retardant known as PBB. Scientists tracking the long-term impact of the toxin on the community found that pregnant women who drank the milk and ate the meat from contaminated cattle gave birth to daughters who started menstruating a year earlier than girls whose mothers were not exposed to PBB.

Parents often fret about giving their children dairy products from cattle raised with bovine growth hormone (rBGH), added to the feed of livestock to increase milk production; they worry that it may contribute to early puberty. There are other reasons to avoid animal products produced by poultry and livestock that consume feed laced with growth hormones; it is likely that feed is laced with antibiotics and that it came from crops sprayed with toxic pesticides, which collect in animal fat, including meat, egg yolks, and milk fat. But there is no conclusive evidence linking rBGH specifically to early puberty in girls.

Still, there is another factor related to our food supply: an overabundance of it.

Diet as a Hormone Hacker

Girls who go through early puberty often have a common denominator — a high body mass index and, corresponding to that, higher levels of body fat. More body fat means more of the hormone leptin, released by fat cells. Leptin is also thought to be instrumental in stimulating the production of estrogen and is necessary for the onset of menstruation. The more fat cells, the more leptin, and the more of the puberty-triggering hormone estrogen. Body fat, breast development, and menstruation are connected.[24]

Here's another tricky thing about leptin: It is also responsible for telling the brain when the body is full or when it's time to eat. Nutritionists point out that an excess of leptin, due in part to too many fat cells, can ultimately "break" this mechanism, meaning that it becomes easier for the brain to ignore leptin's *I'm full!* message, causing overeating. For a little girl headed toward obesity, losing weight and stopping the leptin-estrogen pattern is quite hard. (Conversely, crash dieting and being underweight can also disrupt normal leptin production, leading to delayed or disrupted menstruation, low fertility, and other hormonally dependent functions. Girls who lack adequate body fat — perhaps because of dieting, rigorous athletic activity, or a combination of both — often have late puberty.)

Even if a girl is not overweight, in Western society she is usually quite well nourished from birth. That's not a bad thing. We've survived as a species precisely because we have largely vanquished starvation and many diseases. Ancestral women did not have as many offspring in times of food scarcity, but when conditions were more bountiful, mothers were more bountiful, too (recall "the need to feed versus the need to breed").

But there are consequences of the kind of dense nutrition and high-caloric food we have today. For girls or boys, once the body matures sexually, it is capable of reproduction — even if the mind and spirit are still those of a child. We may inadvertently be resetting our biological clocks to switch into reproductive mode at a younger age. It isn't "early" puberty — it is *earlier* puberty.

We are evolving.

Breaking the Cycle — Temporarily or for Good

Pediatric endocrinologists who treat early-onset puberty in young children may prescribe drugs known as puberty blockers, which stop the release of sex hormones such as estrogen and testosterone and put the brakes on certain physical developments, until a boy or girl "catches up" psychologically to what is happening physically.

More recently, puberty blockers are being used for another, off-label purpose: to treat gender-nonconforming youths who are exploring the transition from one sex to another, and who may just be reaching puberty or are in its early stages. Some transgender advocates, doctors, and mental health professionals support this because gender-nonconforming people may not feel comfortable revealing their status until adulthood, after their physical development is complete. By then it is too late to reverse the full effects of the hormonal cycle without surgery or other serious interventions. In addition, experts—as well as parents of gender-nonconforming children—point out that puberty blockers can alleviate a child's anguish. A girl with gender dysphoria may feel extreme distress at developing breasts or the prospect of menstruation.

The Endocrine Society guidelines suggest that starting at age sixteen, a teen can safely begin cross-sex hormone therapy (different from puberty blockers) if she or he is planning to fully transition surgically, with transgender girls receiving estrogen and transgender boys receiving testosterone—though doctors caution that gender-specific hormones that the system is designed to receive also contribute to brain and bone development. Still, given the rates of depression and suicide among transgender adults, advocates believe the benefits of cross-sex hormone therapy outweigh the risks.

Unlike puberty blockers, cross-sex hormone therapy causes irreversible physical changes, such as the growth of an Adam's apple or facial hair in girls who have testosterone therapy. It's an extremely complicated issue for parents trying to make the right choices. My colleague Eric Vilain, a geneticist and pediatrician, points to a study showing that up to 80 percent of gender dysphoric boys adjust to being male by the time they reach adolescence and do not transition to female when they reach adulthood.[25] (There are no comparable statistics on girls because women—this will sound familiar—have not been studied as closely.) But he also agrees that the emotional pain of living with gender dysphoria can be serious and overwhelming.

To say this is complicated is an understatement, but ultimately it is remarkable that humans have a choice. Should we desire, we can select specific hormones—hormones we were not born with—to permanently alter our sexual and reproductive lives, and redefine our own hormonal intelligence.

Being Hormonally Intelligent

Female estrous behaviors in animals and humans alike were born out of necessity: attract a fit male and reproduce fit offspring. Life was simple, stark, and short. But it had a purpose – survive and thrive. Mate, reproduce, and repeat. Across a vast number of nonhuman species, the ancient behavior, and the desires that underlie it, endures.

Millions of years later, women still experience the same hormonal cycles our female ancestors did, but we have a breathtaking array of choices. In our lifetimes, we can choose one mate, or many, or none at all. We can be with the opposite sex, or our own. We can have children, or not. We can voluntarily alter our hormonal cycles with contraception or hormone therapies, or address unwanted hormonal disruptions through medicine and science. We are modern women who can choose our social and reproductive destinies.

Still, how we choose to listen to and use our hormonal intelligence is complicated – for instance, whom we select as an intimate partner and whether or not to have children (or breastfeed, once they arrive). Long-term relationships and personal choices that change the course of a life are clearly not decisions based only on estrous desires, however strong those desires may be.

In evolutionary psychology and other sciences, the "naturalistic fallacy" means that just because something is "natural" and perhaps instinctive doesn't mean it is "good." This is especially true if the something in question (estrus) is a suite of desires and behaviors designed to meet different challenges from those we face today.

So, what do we do with hormonal intelligence? My perspective is that if we know what's happening in our bodies and our minds, and how it differs across the life course and for us as individuals, we can determine whether to ignore that chocolate-cake-for-breakfast craving or embrace it (because sometimes it tastes *really* good). I hope that hormonally intelligent choices will be informed by the science in this book. Journalists nearly always ask me, What's my advice for women? At first I would bristle at the question (*I am a scientist, not an advice columnist!*). But then I realized I did have something to offer. *Know the science. Know yourself. You will make the most informed decisions.* And isn't that a big part of what science is for?

To make this more concrete, here are some thoughts. Certain estrus-driven behaviors may have served our female ancestors, both human and nonhuman, quite well, such as attracting and mating with an alpha male and bearing his offspring, or being fiercely competitive with other females in order to protect the life of those offspring. Evolution is about outreproducing the competition, after all. A twenty-first-century woman may experience those same hormonal nudges, but does she necessarily benefit from a compulsion to seek out an alpha male for sex if she is in a committed relationship? It's unlikely. What if she acts on an urge to go head-to-head with a female colleague and sabotages her efforts at work? That could backfire. Ornamentation and roaming behaviors can take the form of showing skin, flirting, and going on a bar crawl — but those are not the strategic behaviors we might think are best to choose if there is a valued spouse or partner, work or school the next day, or children waiting at home.

On the other hand, looking good and feeling strong, going out and meeting new people, and knowing what — and whom — to avoid are strategic behaviors we might embrace, and hormonal ones, as well. So is the urge to protect and defend a baby, or find a loving, helpful, and steadfast partner. Or maybe a well-timed fling is just what we want. We have the power to choose — and to put our choices in the context of our hormonal life.

Women can think logically and make rational decisions every day (as well as make mistakes, or be biased — sometimes in a good way). This is because we are not under strict hormonal control, locked in the sway of "heat," weakened by the loss of blood, or depleted as our fertility fades. Still, when we do feel these ancient forces stirring in rhythm with our hormonal cycles, we can tap into a uniquely female power.

In my view, every girl and woman benefits from understanding the scope of hormonal cycles, the hows, whens, and whys. We should become familiar with the potential nudges that affect our behavior. And we should know that choosing to act on those behaviors is an individual choice, dependent upon our own preferences and goals. Being naive to our hormonal natures will not help us. Being hormonally intelligent, on the other hand, will.

It took too long for those of us in the scientific community to admit that human estrus is real. Now we are making up for lost time as we seek to research and understand its implications. All women will benefit if we are better educated about our female bodies and minds. And we will benefit if men are more knowledgeable as well.

She's hormonal.

The next time you hear — or say — those words, consider that "she" is a grandmother, a mother, a sister, a friend, a daughter. "She" is one in the unbroken chain of women who were our ancestors living eons ago up through the present, and who are yet to be born and come of age, each one possessing a singular hormonal cycle. "She" may be you.

"She" is me, and I'm proud to be hormonal.

Acknowledgments

I am grateful for my many mentors, academic and personal, who helped me find my way to writing this book, beginning with Dr. Dan Moriarty, who taught me as an undergraduate. I was fascinated by human behavior and eager to explain what I saw, but I was also unimpressed with the theories of the time, which left little room for biology. Dan taught me psychology, behavioral genetics, sexuality (in animals, peppered with tantalizing hints about humans), and a little about hormones too. He fit the stereotype of the stern professor: bearded, stacking and restacking his notes on the podium, rarely cracking a smile during lectures, and offering his classes only at eight a.m. But he was a softie underneath that gruff exterior. Students were either half asleep or too timid to ask him questions. But one day he introduced parental investment theory as a way of explaining the differences in mating behavior between male and female squirrels. I thought: But this explains a lot about human behavior too. I see it every weekend! I raised my hand and asked if anyone was using this theory to explain human sex differences.

Dan told me that one person was: Dr. David Buss at the University of Michigan. The name David Buss was seared into my brain, though it would be years before our paths finally crossed. Even so, I felt a new focus on my future. This was what I wanted to do!

I completed my master's degree at my beloved William & Mary, where I met Dr. Lee Kirkpatrick, a hippy intellectual with a long gray ponytail who drove a VW bus and studied interpersonal attachment (including our imagined attachment to God). He took me to Grateful Dead shows — and, fatefully, to a talk given by Dr. Buss at a local university. The room was packed with several hundred attendees. Although I felt like an interloper in the Q&A, I raised my hand and asked him how our ancient psychology plays out in the modern world. What were the implications, I asked.... He pointed at me and told me... that is a "brilliant question!" It was not really a brilliant question, but his words encouraged me to investigate the ideas that would eventually become this book.

I went on to pursue my PhD with David. He has been a generous mentor and friend, and I am fortunate to have found him as an intellectual match. We would have marathon meetings — two hours or more in length — just talking about ideas. It was a dream come true.

I followed David to the University of Texas, where I completed my degree. I had lovely graduate-student colleagues: April Bleske, who was my office mate; Lisa Redford; Sergei Bogdanov; Todd Shackelford; and, yes, even you, Barry (Freidman). I was lucky to have great professors, including Randy Deihl, Devendra Singh, Arnie Buss, Michael Ryan, and Cindy Meston.

At UCLA, where I landed after completing my PhD, I met Anne Peplau and Christine Dunkel Schetter, two of the most supportive and inspiring academics I know. Anne was among the first to study female sexual orientation (quantifying it and surveying women about their experiences). She was a bold pioneer — and a role model to me. Chris amazes me, working on really difficult problems, such as figuring out what happens in pregnancy that results in the most common, debilitating problem in pregnancy:

preterm birth. She can secure three NIH grants at once, something I've never seen anyone else do.

I also have other senior colleagues who generously offered their advice and inspiration: Joan Silk, Leda Cosmides, Jim Sidanius, Art Arnold, David Sears, Michael Bailey, Don Symons, Randy Thornhill, and John Tooby, among others. At more of a distance, I have been inspired by other Darwinian feminists (who I hope will not mind my applying this moniker to them), including Sarah Hrdy and Patty Gowaty. I am grateful to my dear friend Beth Schumann, who workshopped my early pages and encouraged me with her excitement about the project.

I have had wonderful peers at UCLA: Abigail Saguy, Clark Barrett, Greg Bryant, Eric Vilain, Naomi Eisenberger, Dan Blumstein, and Dan Fessler (by whom we get "Fessled," which means after a talk, we get asked a tough question in one "Dan-o-second"), among many others. And I'm grateful to many others who have been there for me intellectually and personally: Debra Lieberman and Kerri Johnson (who my daughter would say are my BFFs), Bill von Hippel, Athena Aktipis, Jeffrey Sherman, Doug Kenrick, Steve Neuberg, Daniel Nettle, and Dominic Johnson. All of you have been absolutely lovely as colleagues, supporters, and friends. Geoffrey Miller, who is a brilliant science writer, helped me conceive the idea for this book and then pushed me to write it. Thank you.

Steve Gangestad, whose work takes center stage in the chapter on "heat seeking," has been like a second PhD mentor to me. He's a brilliant statistician, methodologist, and idea maker. When I have faced political opposition to my work, he has been there to listen (and share a little outrage). I have learned a huge amount from him. Many of the ideas in this book were born out of discussions I had with him over the years. I would not have written this book if it were not for Steve's work, our collaborations, and his friendship.

Most important are my graduate students. Elizabeth Pillsworth, my first student from my years at UCLA, helped me develop research methods that we still use in the lab today. Much of the research in this book would not have happened without her keen insight. Many other students followed in her path: David Frederick, Josh Poore, Shimon Saphire-Bernstein, Andrew Galperin, Christina Larson, Kelly Gildersleeve, Melissa Fales, Britt Ahlstrom, David Pinsof, Jessica Shropshire, and Tran Dinh. I have also been fortunate to have whip-smart postdocs in my lab: Jennifer Hahn-Holbrook, Damian Murray, and Aaron Lukaszewski. Jen inspired much of what I wrote about in Chapter 7, "Maidens to Matriarchs." All of these students opened my eyes to new ideas and made my job the best on the planet. They are family. Amanda Barnes helped me immensely as a research assistant. I could not have made it through the homestretch without her.

And I am grateful to Katinka Matson, my extraordinarily patient agent, who helped me hone my ideas for this book and who guided it straight into the hands of my wise, clear-eyed (and also extraordinarily patient) editor, Tracy Behar. Thanks to both of you for sticking by me. Becky Cabaza, my "writing partner," whom I have never even laid eyes on, you are another BFF. Thank you for helping me translate my ideas into words for all of the "hormonal women" out there to read. And thank you for being a ton of fun to talk to and to bounce ideas off of, and for being a source of immense support, professionally and personally.

Thanks to all of you. Your work helped me create the book you are holding today, and I'd like to think that together, we're all contributing to deepening our collective hormonal intelligence.

Notes

Introduction: *The New Darwinian Feminism*

1. Gloria Steinem, "If Men Could Menstruate," in *Outrageous Acts and Everyday Rebellions* (New York: NAL, 1986), posted by Sally Kohn, http://ww3.haverford.edu/psychology/ddavis/p109g/steinem.menstruate.html.

Chapter 1: *The Trouble with Hormones*

1. Claudia Goldin, Lawrence F. Katz, and Ilyana Kuziemko, "The Homecoming of American College Women: The Reversal of the College Gender Gap," *Journal of Economic Perspectives* 20, no. 4 (2006): 133–156.

2. Kristina M. Durante, Ashley Rae, and Vladas Griskevicius, "The Fluctuating Female Vote: Politics, Religion, and the Ovulatory Cycle," *Psychological Science* 24, no. 6 (2013): 1007–1016.

3. Katie Baker, "CNN Thinks Crazy Ladies Can't Help Voting with Their Vaginas Instead of Their Brains," *Jezebel*, October 24, 2012, http://jezebel.com/5954617/cnn-thinks-crazy-ladies-cant-help-voting-with-their-vaginas-instead-of-their-brains; Kate Clancy, "Hot for Obama, but Only When This Smug Married Is Not Ovulating," *Scientific American*, October 26, 2012, https://blogs.scientificamerican.com/context-and-variation/hot-for-obama-ovulation-politics-women/; Alexandra Petri, "CNN's Hormonal Lady Voters," *Washington Post*, October 25, 2012, https://www.washingtonpost.com/blogs/compost/post/cnns-hormonal-lady-voters/2012/10/24/961799c4-1e1f-11e2-9cd5-b55c38388962_blog.html?utm_term=.48f969c61461.

4. Marylin Bender, "Doctors Deny Woman's Hormones Affect Her as an Executive," *New York Times*, July 31, 1970.

5. Nancy Ross, "Berman Says He Won't Quit," *Washington Post, Times Herald*, July 31, 1970.

6. "History," Our Bodies Ourselves, http://www.ourbodiesourselves.org/history/.

7. Jayne Riew, *The Invisible Month*, http://theinvisiblemonth.com/.

8. Jayne Riew, "The Artist," *The Invisible Month*, http://theinvisible month.com/.

9. Martie G. Haselton and Steven W. Gangestad, "Conditional Expression of Women's Desires and Men's Mate Guarding across the Ovulatory Cycle," *Hormones and Behavior* 49 (2006) 509–518; Martie G. Haselton and Kelly Gildersleeve, "Human Ovulation Cues," *Current Opinion in Psychology* 7 (2016): 120–125.

10. "Policy & Compliance," National Institutes of Health, http://grants.nih.gov/grants/policy/policy.htm.

11. G. H. Wang, "The Relation between 'Spontaneous' Activity and the Oestrous Cycle in the White Rat," *Comparative Psychology Monographs* 6 (1923): 1–40.

12. Malin Ah-King, Andrew B. Barron, and Marie E. Herberstein, "Genital Evolution: Why Are Females Still Understudied?" *PLoS Biology* 12: e1001851, doi: 10.1371/journal.pbio.1001851.

13. Ibid.

14. Patricia L. R. Brennan, Richard O. Prum, Kevin G. McCracken, Michael D. Sorenson, Robert E. Wilson, and Tim R. Birkhead, "Coevolution of Male and Female Genital Morphology in Waterfowl," *PLoS One* 2: e418, doi: 10.1371/journal.pone.0000418.

Chapter 2: Heat Seekers

1. Steven W. Gangestad and Randy Thornhill, "Menstrual Cycle Variation in Women's Preferences for the Scent of Symmetrical Men," *Proceedings of the Royal Society B: Biological Sciences* 265 (1998): 927–933.

2. Cues of "good genes" might have had huge consequences ancestrally, even if they have little consequence today. We live in an era of modern medicine and abundant food, and most of us have a generally cushy lifestyle. So, here and throughout the book, when I refer to "cues" or "indicators" of good genes, I refer to cues that ancestral women used to choose high-quality mates. They don't necessarily give women's offspring benefits in the modern world.

3. Randy J. Nelson, *An Introduction to Behavioral Endocrinology*, 3rd ed. (Sunderland, MA: Sinauer Associates, 2005).

4. Alan F. Dixson, *Primate Sexuality: Comparative Studies of Prosimians, Monkeys, Apes, and Human Beings*, 2nd ed. (Oxford: Oxford University Press, 2012).

5. Owen R. Floody and Donald W. Pfaff, "Aggressive Behavior in Female Hamsters: The Hormonal Basis for Fluctuations in Female Aggressiveness Correlated with Estrous State," *Journal of Comparative and Physiological Psychology* 91 (1977): 443–464.

6. Ibid.; Nelson, *Introduction to Behavioral Endocrinology.*

7. Carol Diakow, "Motion Picture Analysis of Rat Mating Behavior," *Journal of Comparative and Physiological Psychology* 88 (1975): 704–712; Donald W. Pfaff, Carol Diakow, Michael Montgomery, and Farish A. Jenkins, "X-Ray Cinematographic Analysis of Lordosis in Female Rats," *Journal of Comparative and Physiological Psychology* 92 (1978): 937–941.

8. Dixson, *Primate Sexuality.*

9. Ibid.; Nelson, *Introduction to Behavioral Endocrinology.*

10. Mark Griffith, *Aeschylus: Prometheus Bound* (Cambridge: Cambridge University Press, 1983).

11. Plato, *The Republic and Other Works,* trans. Benjamin Jowett (New York: Anchor Books, 1973).

12. Homer, *The Odyssey,* trans. Robert Fagles (New York: Penguin Books, 1996).

13. Jeremiah 2:24 (GNT).

14. Nelson, *Introduction to Behavioral Endocrinology.*

15. Dixson, *Primate Sexuality.*

16. P. G. McDonald and Bengt J. Meyerson, "The Effect of Oestradiol, Testosterone and Dihydrotestosterone on Sexual Motivation in the Ovariectomized Female Rat," *Physiology and Behavior* 11 (1973): 515–520; Bengt J. Meyerson, Leif Lindström, Erna-Britt Nordström, and Anders Ågmo, "Sexual Motivation in the Female Rat after Testosterone Treatment," *Physiology and Behavior* 11 (1973): 421–428.

17. Frank Beach, "Locks and Beagles," *American Psychologist* 24 (1969): 971–989.

18. Ibid.

19. Frank Beach, "Sexual Attractivity, Proceptivity, and Receptivity in Female Mammals," *Hormones and Behavior* 7 (1976): 105–138.

20. See Chapter 1 for more information on how this disparity persists to this day, and how it's being addressed by groups like the NIH.

21. Beach, "Locks and Beagles." Beach described the change in his thinking in a keynote address delivered to the American Psychological Association — the largest professional association of psychologists at the time. With characteristic good humor, he titled his talk "Locks and Beagles."

22. Martha K. McClintock, "Sociobiology of Reproduction in the Norway Rat (*Rattus norvegicus*): Estrous Synchrony and the Role of the Female Rat in Copulatory Behavior" (PhD diss., ProQuest Information and Learning, 1975).

23. Martha K. McClintock and Norman T. Adler, "The Role of the Female during Copulation in Wild and Domestic Norway Rats (*Rattus norvegicus*)," *Behaviour* 67 (1978): 67–96.

24. Ibid.; Mary S. Erskine, "Solicitation Behavior in the Estrous Female Rat: A Review," *Hormones and Behavior* 23 (1989): 473–502.

25. Martha K. McClintock, "Group Mating in the Domestic Rat as a Context for Sexual Selection: Consequences for the Analysis of Sexual Behavior and Neuroendocrine Responses," *Advances in the Study of Behavior* 14 (1984): 1–50.

26. Also see the fascinating discussion of pleasure in rats and choice on the basis of "good sex" (from the rat perspective) in chapters 3 and 14 of *Vagina: A New Biography* (New York: Ecco, 2012).

27. Simona Cafazzo, Roberto Bonanni, Paola Valsecchi, and Eugenia Natoli, "Social Variables Affecting Mate Preferences, Copulation and Reproductive Outcome in a Pack of Free-Ranging Dogs," *PLoS One* 6 (2014): e98594, doi: 10.1371/journal.pone.0098594.

28. Akiko Matsumoto-Oda, "Female Choice in the Opportunistic Mating of Wild Chimpanzees (*Pan troglodytes schweinfurthii*) at Mahale," *Behavioral Ecology and Sociobiology* 46 (1999): 258–266. But see Rebecca M. Stumpf and Cristophe Boesch, "Does Promiscuous Mating Preclude Female Choice? Female Sexual Strategies in Chimpanzees (*Pan troglodytes verus*) of the Taï National Park, Côte d'Ivoire," *Behavioral Ecology and Sociobiology* 57 (2005): 511–524. The latter showed that females in one group mated with both high- and low-ranking males but not middle-ranking males when fertility was at its highest point. It is possible that females get genetic benefits from the high-ranking males and nongenetic benefits (e.g., access to food or protection from low-ranking males in exchange for mating with them when fertile).

29. Ekaterina Klinkova, J. Keith Hodges, Kerstin Fuhrmann, Tom de Jong, and Michael Heistermann, "Male Dominance Rank, Female Mate Choice and Male Mating and Reproductive Success in Captive Chimpanzees," *International Journal of Primatology* 26 (2005): 357–384.

30. Pascal R. Marty, Maria A. Van Noordwijk, Michael Heistermann, Erik P. Willems, Lynda P. Dunkel, Manuela Cadilek, Muhammad Agil, and Tony Weingrill, "Endocrinological Correlates of Male Bimaturism in Wild Bornean Orangutans," *American Journal of Primatology* 77, no. 11 (2015): 1170–1178.

31. Cheryl D. Knott, Melissa E. Thompson, Rebecca M. Stumpf, and Matthew H. McIntyre, "Female Reproductive Strategies in Orangutans, Evidence for Female Choice and Counterstrategies to Infanticide in a Species with Frequent Sexual Coercion," *Proceedings of the Royal Society B: Biological Sciences* 277 (2010): 105–113; Parry M. R. Clarke, S. Peter Henzi, and Louise Barrett, "Sexual Conflict in Chacma Baboons, *Papio hamadryas ursinus*: Absent Males Select for Proactive Females," *Animal Behaviour* 77 (2009): 1217–1225. The evidence here is somewhat hard to interpret because dominant

males might also forcibly exclude subordinate males. There's also some evidence that many primates spread around their matings to confuse paternity.

32. Tony Weingrill, John E. Lycett, and S. Peter Henzi, "Consortship and Mating Success in Chacma Baboons (*Papio hamadruas ursinus*)," *Ethology* 106 (2000): 1033–1044.

33. Charles Darwin, *The Descent of Man, and Selection in Relation to Sex* (London: J. Murray, 1871).

34. George W. Corner, *The Hormones in Human Reproduction* (Princeton, NJ: Princeton University Press, 1942).

35. Nelson, *Introduction to Behavioral Endocrinology.*

36. Allen J. Wilcox, Clarice R. Weinberg, and Donna D. Baird, "Timing of Sexual Intercourse in Relation to Ovulation: Effects on the Probability of Conception, Survival of the Pregnancy, and Sex of the Baby," *New England Journal of Medicine* 333 (1995): 1517–1521.

37. J. Richard Udry and Naomi M. Morris, "Distribution of Coitus in the Menstrual Cycle," *Nature* 220 (1968): 593–596.

38. Allen J. Wilcox, Donna D. Baird, David B. Dunson, Robert McConnaughey, James S. Kesner, and Clarice R. Weinberg, "On the Frequency of Intercourse around Ovulation: Evidence for Biological Influences," *Human Reproduction* 19 (2004): 1539–1543.

39. David A. Adams, Alice R. Gold, and Anne D. Burt, "Rise in Female-Initiated Sexual Activity at Ovulation and Its Suppression by Oral Contraceptives," *New England Journal of Medicine* 299 (1978): 1145–1150; Susan B. Bullivant, Sarah A. Sellergren, Kathleen Stern, Natasha A. Spencer, Suma Jacob, Julie A. Mennella, and Martha K. McClintock, "Women's Sexual Experience during the Menstrual Cycle: Identification of the Sexual Phase by Noninvasive Measurement of Luteinizing Hormone," *Journal of Sex Research* 41 (2004): 82–93.

40. S. Marie Harvey, "Female Sexual Behavior: Fluctuations during the Menstrual Cycle," *Journal of Psychosomatic Research* 31 (1987): 101–110.

41. Bullivant et al., "Women's Sexual Experience."

42. Alexandra Brewis and Mary Meyer, "Demographic Evidence That Human Ovulation Is Undetectable (at Least in Pair Bonds)," *Current Anthropology* 46 (2005): 465–471.

43. Ibid.

44. Pamela C. Regan, "Rhythms of Desire: The Association between Menstrual Cycle Phases and Female Sexual Desire," *Canadian Journal of Human Sexuality* 5 (1996): 145–156.

45. Martie G. Haselton and Steven W. Gangestad, "Conditional Expression of Women's Desires and Men's Mate Guarding across the Ovulatory Cycle," *Hormones and Behavior* 49 (2006): 509–518; Christina M. Larson,

"Do Hormonal Contraceptives Alter Mate Choice and Relationship Functioning in Humans?" (PhD diss., University of California, Los Angeles, 2014); Steven W. Gangestad, Randy Thornhill, and Christine E. Garver, "Changes in Women's Sexual Interests and Their Partner's Mate-Retention Tactics across the Ovulatory Cycle: Evidence for Shifting Conflicts of Interest," *Proceedings of the Royal Society B: Biological Sciences* 269 (2002): 975–982.

46. James R. Roney and Zach L. Simmons, "Hormonal Predictors of Sexual Motivation in Natural Menstrual Cycles," *Hormones and Behavior* 63 (2013): 636–645.

47. J. Richard Udry and Naomi M. Morris, "Variations in Pedometer Activity during the Menstrual Cycle," *Obstetrics and Gynecology* 35 (1970): 199–201.

48. Richard L. Doty, M. Ford, George Preti, and G. R. Huggins, "Changes in the Intensity and Pleasantness of Human Vaginal Odors during the Menstrual Cycle," *Science* 190 (1975): 1316–1318.

49. All of the ratings were on the unattractive end of the scale, so to be more technically correct, it's better to say that the high-fertility samples were rated as less unattractive. The study was conducted in the seventies, a heyday of feminine "hygiene" products. I would guess that many people have more evolved sensibilities about body odors now, but the recent "Big Data" book, *Everybody Lies*, showed that one of the most frequent Google searches for women concerns whether their vaginal odor smells bad. Apparently these searches are dominated by younger women, so perhaps there's a wisdom of embracing our hormonal natures that comes with a little more sexual experience. Seth Stephens-Davidowitz, *Everybody Lies: Big Data, New Data, and What the Internet Can Tell Us about Who We Really Are* (New York: Harper Collins, 2017).

50. Steven Pinker, *The Blank Slate* (New York: Penguin Books, 2002). In this book, Pinker does not endorse these ideas, but he describes them in detail, along with their origins and pitfalls.

51. Robert Trivers, *Parental Investment and Sexual Selection*, vol. 136 (Cambridge, MA: Biological Laboratories, Harvard University, 1972).

52. Ibid.

53. Ibid.

54. Terri D. Conley, Amy C. Moors, Jes L. Matsick, Ali Ziegler, and Brandon A. Valentine, "Women, Men, and the Bedroom: Methodological and Conceptual Insights That Narrow, Reframe, and Eliminate Gender Differences in Sexuality," *Current Directions in Psychological Science* 20 (2011): 296–300; David P. Schmitt, Peter K. Jonason, Garrett J. Byerley, Sandy D. Flores, Brittany E. Illbeck, Kimberly N. O'Leary, and Ayesha Qudrat, "A

Reexamination of Sex Differences in Sexuality: New Studies Reveal Old Truths," *Current Directions in Psychological Science* 21 (2012): 135–139.

55. Russell D. Clark and Elaine Hatfield, "Gender Differences in Receptivity to Sexual Offers," *Journal of Psychology and Human Sexuality* 2, no. 1 (1989): 39–55.

56. D. P. Schmitt, L. Alcalay, J. Allik, L. Ault, I. Austers, K. L. Bennett, G. Bianchi, et al., "Universal Sex Differences in the Desire for Sexual Variety: Tests from 52 Nations, 6 Continents, and 13 Islands," *Journal of Personality and Social Psychology* 85 (2003): 85–104; David Schmidt, "Fundamentals of Human Mating Strategies," in *The Handbook of Evolutionary Psychology*, ed. David Buss (Hoboken, NJ: John Wiley and Sons, 2016), 294–316, http://www.wiley.com/WileyCDA/WileyTitle/productCd-111875588X.html.

57. David M. Buss and David P. Schmitt, "Sexual Strategies Theory: An Evolutionary Perspective on Human Mating," *Psychological Review* 100 (1993): 204–232.

58. Schmitt et al., "Universal Sex Differences."

59. Buss and Schmitt, "Sexual Strategies Theory."

60. Trivers, *Parental Investment*; Randy Thornhill and Steven W. Gangestad, *The Evolutionary Biology of Human Female Sexuality* (New York: Oxford University Press, 2008); Anders P. Moller and Randy Thornhill, "Bilateral Symmetry and Sexual Selection: A Meta-Analysis," *American Naturalist* 151 (1998): 174–192.

61. Steven W. Gangestad and Jeffry A. Simpson, "The Evolution of Human Mating: Trade-Offs and Strategic Pluralism," *Behavioral and Brain Sciences* 23 (2000): 573–587.

62. Gangestad and Thornhill, "Menstrual Cycle Variation."

63. Ibid.

64. Ibid.

65. Ian S. Penton-Voak and David I. Perrett, "Female Preference for Male Faces Changes Cyclically: Further Evidence," *Evolution and Human Behavior* 21 (2000): 39–48.

66. Kelly Gildersleeve, Martie G. Haselton, and Melissa R. Fales, "Do Women's Mate Preferences Change across the Ovulatory Cycle? A Meta-Analytic Review," *Psychological Bulletin* 140, no. 5 (2014): 1205.

67. Anja Rikowski and Karl Grammer, "Human Body Odour, Symmetry, and Attractiveness," *Proceedings of the Royal Society B: Biological Sciences* 266 (1999): 869–874; Penton-Voak and Perrett, "Female Preference"; Victor S. Johnston, Rebecca Hagel, Melissa Franklin, Bernhard Fink, and Karl Grammer, "Male Facial Attractiveness: Evidence for Hormone-Mediated Adaptive Design," *Evolution and Human Behavior* 22 (2001): 251–267; Randy Thornhill, Steven W. Gangestad, Robert Miller, Glenn Scheyd, Julie K.

McCollough, and Melissa Franklin, "Major Histocompatibility Complex Genes, Symmetry, and Body Scent Attractiveness in Men and Women," *Behavioral Ecology* 14 (2003): 668–678; Randy Thornhill and Steven W. Gangestad, "The Scent of Symmetry: A Human Sex Pheromone That Signals Fitness?" *Evolution and Human Behavior* 20 (1999): 175–201.

Chapter 3: Around the Moon in Twenty-Eight Days

1. This issue concerning terminology is actually one of surprisingly bitter controversy (surprising to me, and perhaps to you!). Alan Dixson is regarded as the world expert on primate sexuality, having written a book called, fittingly, *Primate Sexuality*, an encyclopedic, exceptionally scholarly tome containing nearly three thousand references, to which I refer often in my own work (and here in this book on many occasions). I trust and greatly respect his scholarship on nonhuman primates. When it comes to our species, he does not seem very open to the abundant evidence for an estrus-like state that has accumulated in the last two decades. His book was published in 2012, well after research on human estrus had taken flight (and humans are, of course, primates). He bridles at the use of the term "estrus" with respect to humans and argues instead that the only appropriate term is "the menstrual cycle." He would like estrus to be reserved for species in which sexual behavior is limited only to the fertile window of the cycle. However, I agree with my colleagues Steve Gangestad and Randy Thornhill that even if human sexuality appears to be more flexible than in species with "classic estrus," there is also a great deal of evidence for estrus-like changes — women's sexual desires and mating-related behaviors. Using a unique human term walls off the human case in ways that could prevent us from exploring parallels with our nonhuman cousins. To see how acrimonious the discussion can get, you can check out Dixson's review (https://www.amazon.com/dp/019534099X/ref=rdr_ext_tmb) of Thornhill and Gangestad's 2008 book: Randy Thornhill and Steven W. Gangestad, *The Evolutionary Biology of Human Female Sexuality* (New York: Oxford University Press, 2008).

2. James R. Roney and Zachary L. Simmons, "Elevated Psychological Stress Predicts Reduced Estradiol Concentrations in Young Women," *Adaptive Human Behavior and Physiology* 1, no. 1 (2015): 30–40; Samuel K. Wasser and David P. Barash, "Reproductive Suppression among Female Mammals: Implications for Biomedicine and Sexual Selection Theory," *Quarterly Review of Biology* 58, no. 4 (1983): 513–538; Samuel K. Wasser, "Psychosocial Stress and Infertility," *Human Nature* 5, no. 3 (1994): 293–306.

3. Gordon D. Niswender, Jennifer L. Juengel, Patrick J. Silva, M. Keith Rollyson, and Eric W. McIntush, "Mechanisms Controlling the Function and

Life Span of the Corpus Luteum," *Physiological Reviews* 80, no. 1 (2000): 1–29.

4. Martha K. McClintock, "Menstrual Synchrony and Suppression," *Nature* (1971).

5. Beverly I. Strassmann, "Menstrual Synchrony Pheromones: Cause for Doubt," *Human Reproduction* 14, no. 3 (1999): 579–580.

6. Julia Ostner, Charles L. Nunn, and Oliver Schülkea, "Female Reproductive Synchrony Predicts Skewed Paternity across Primates," *Behavioral Ecology* 19, no. 6 (2008): 1150–1158.

7. Raymond Greene and Katharina Dalton, "The Premenstrual Syndrome," *British Medical Journal* 1, no. 4818 (1953): 1007.

8. M. J. Law Smith, David I. Perrett, Benedict C. Jones, R. Elisabeth Cornwell, Fhionna R. Moore, David R. Feinberg, Lynda G. Boothroyd, et al., "Facial Appearance Is a Cue to Oestrogen Levels in Women," *Proceedings of the Royal Society B: Biological Sciences* 273, no. 1583 (2006): 135–140.

9. Kristina M. Durante and Norman P. Li, "Oestradiol Level and Opportunistic Mating in Women," *Biology Letters* 5, no. 2 (2009): 179–182.

10. Grazyna Jasieńska, Anna Ziomkiewicz, Peter T. Ellison, Susan F. Lipson, and Inger Thune, "Large Breasts and Narrow Waists Indicate High Reproductive Potential in Women," *Proceedings of the Royal Society B: Biological Sciences* 271, no. 1545 (2004): 1213.

11. James R. Roney and Zachary L. Simmons, "Women's Estradiol Predicts Preference for Facial Cues of Men's Testosterone," *Hormones and Behavior* 53, no. 1 (2008): 14–19.

12. Durante and Li, "Oestradiol Level."

13. Steven J. Stanton and Oliver C. Schultheiss, "Basal and Dynamic Relationships between Implicit Power Motivation and Estradiol in Women," *Hormones and Behavior* 52, no. 5 (2007): 571–580; Steven J. Stanton and Robin S. Edelstein, "The Physiology of Women's Power Motive: Implicit Power Motivation Is Positively Associated with Estradiol Levels in Women," *Journal of Research in Personality* 43, no. 6 (2009): 1109–1113.

14. Lebron-Milad Kelimer, Bronwyn M. Graham, and Mohammed R. Milad, "Low Estradiol Levels: A Vulnerability Factor for the Development of Posttraumatic Stress Disorder," *Biological Psychiatry* 72, no. 1 (2012): 6–7.

15. J. Richard Udry and Naomi M. Morris, "Variations in Pedometer Activity during the Menstrual Cycle," *Obstetrics and Gynecology* 35 (1970): 199–201.

16. James R. Roney and Zach L. Simmons, "Hormonal Predictors of Sexual Motivation in Natural Menstrual Cycles," *Hormones and Behavior* 63 (2013): 636–645.

17. Dionne P. Robinson and Sabra L. Klein, "Pregnancy and Pregnancy-Associated Hormones Alter Immune Responses and Disease Pathogenesis," *Hormones and Behavior* 62, no. 3 (2012): 263–271.

18. Diana S. Fleischman and Daniel M. T. Fessler, "Progesterone's Effects on the Psychology of Disease Avoidance: Support for the Compensatory Behavioral Prophylaxis Hypothesis," *Hormones and Behavior* 59, no. 2 (2011): 271–275.

19. Monika Østensen, Peter M. Villiger, and Frauke Förger, "Interaction of Pregnancy and Autoimmune Rheumatic Disease," *Autoimmunity Reviews* 11, no. 6 (2012): A437–A446.

20. Fleischman and Fessler, "Progesterone's Effects."

21. Smith et al., "Facial Appearance."

22. Fleischman and Fessler, "Progesterone's Effects."

23. Jon K. Maner and Saul L. Miller, "Hormones and Social Monitoring: Menstrual Cycle Shifts in Progesterone Underlie Women's Sensitivity to Social Information," *Evolution and Human Behavior* 35, no. 1 (2014): 9–16.

24. E. M. Seidel, G. Silani, H. Metzler, H. Thaler, C. Lammb, R. C. Gur, I. Kryspin-Exner, U. Habel, and B. Derntl, "The Impact of Social Exclusion vs. Inclusion on Subjective and Hormonal Reactions in Females and Males," *Psychoneuroendocrinology* 38 (2013): 2925–2932.

25. Oliver C. Schultheiss, Anja Dargel, and Wolfgang Rohde, "Implicit Motives and Gonadal Steroid Hormones: Effects of Menstrual Cycle Phase, Oral Contraceptive Use, and Relationship Status," *Hormones and Behavior* 43, no. 2 (2003): 293–301.

26. Erika Timby, Matts Balgård, Sigrid Nyberg, Olav Spigset, Agneta Andersson, Joanna Porankiewicz-Asplund, Robert H. Purdy, Di Zhu, Torbjörn Bäckström, and Inger Sundström Poromaa, "Pharmacokinetic and Behavioral Effects of Allopregnanolone in Healthy Women," *Psychopharmacology* 186, no. 3 (2006): 414.

27. April Smith, Saul Miller, Lindsay Bodell, Jessica Ribeiro, Thomas Joiner Jr., and Jon Maner, "Cycles of Risk: Associations between Menstrual Cycle and Suicidal Ideation among Women," *Personality and Individual Differences* 74 (2015): 35–40.

28. Sigrid Nyberg, Torbjörn Bäckström, Elisabeth Zingmark, Robert H. Purdy, and Inger Sundström Poromaa, "Allopregnanolone Decrease with Symptom Improvement during Placebo and Gonadotropin-Releasing Hormone Agonist Treatment in Women with Severe Premenstrual Syndrome," *Gynecological Endocrinology* 23, no. 5 (2007): 257–266.

29. Anahad O'Connor, "Katharina Dalton, Expert on PMS, Dies at 87," *New York Times*, October 28, 2010, http://www.nytimes.com/2004/09/28/science/katharina-dalton-expert-on-pms-dies-at-87.html.

30. One of the brilliant founders of the field of evolutionary psychology, Leda Cosmides, shared this idea with me once over dinner. It has since been discussed in the literature, but to my knowledge, she was the first to propose it.

31. Bill de Blasio and Julie Menin, "From Cradle to Cane: The Cost of Being a Female Consumer," New York City Department of Consumer Affairs, December 2015, https://www1.nyc.gov/assets/dca/downloads/pdf/partners/Study-of-Gender-Pricing-in-NYC.pdf.

32. Free the Tampons, http://www.freethetampons.org/.

33. Mike Martin, "The Mysterious Case of the Vanishing Genius," *Psychology Today,* May 1, 2012, https://www.psychologytoday.com/articles/201204/the-mysterious-case-the-vanishing-genius.

34. Deena Emera, Roberto Romero, and Günter Wagner, "The Evolution of Menstruation: A New Model for Genetic Assimilation," *Bioessays* 34, no. 1 (2012): 26–35.

35. Beverly I. Strassmann, "The Evolution of Endometrial Cycles and Menstruation," *Quarterly Review of Biology* 71, no. 2 (1996): 181–220.

Chapter 4: The Evolution of Desire

1. David M. Buss, *The Evolution of Desire,* rev. ed. (New York: Basic Books, 2008).

2. Randy Thornhill and Steven W. Gangestad, *The Evolutionary Biology of Human Female Sexuality* (New York: Oxford University Press, 2008), 286–320.

3. Steven W. Gangestad and Martie G. Haselton, "Human Estrus: Implications for Relationship Science," *Current Opinion in Psychology* 1 (2015): 45–51.

4. E. G. Pillsworth and M. G. Haselton, "Women's Sexual Strategies: The Evolution of Long-Term Bonds and Extra-Pair Sex," *Annual Review of Sex Research* 17 (2006): 59–100.

5. Karin Isler and Carel P. Van Schaik, "How Our Ancestors Broke through the Gray Ceiling: Comparative Evidence for Cooperative Breeding in Early Homo," *Current Anthropology* 53, no. S6 (2012): S453–S465.

6. Richard Wrangham, *Catching Fire: How Cooking Made Us Human* (New York: Basic Books, 2009).

7. "The Teen Brain Still Under Construction," National Institute of Mental Health, https://www.nimh.nih.gov/health/publications/the-teen-brain-6-things-to-know/index.shtml.

8. D. D. Clark and L. Sokoloff, "Circulation and Energy Metabolism of the Brain," in *Basic Neurochemistry: Molecular, Cellular and Medical Aspects,*

ed. G. J. Siegel, B. W. Agranoff, R. W. Albers, S. K. Fisher, and M. D. Uhler (Philadelphia: Lippincott, 1999), 637–670.

9. For a wickedly smart and culture-conscious dissection of "good boys" and "bad boys" and women's sexual pleasure, see chapter 14 in Naomi Wolf's *Vagina: A New Biography* (New York: Ecco, 2012.)

10. Nicholas M. Grebe, Steven W. Gangestad, Christine E. Garver-Apgar, and Randy Thornhill, "Women's Luteal-Phase Sexual Proceptivity and the Functions of Extended Sexuality," *Psychological Science* 24, no. 10 (2013): 2106–2110.

11. Martie G. Haselton and David M. Buss, "Error Management Theory: A New Perspective on Biases in Cross-Sex Mind Reading," *Journal of Personality and Social Psychology* 78, no. 1 (2000): 81–91.

12. Katharina C. Engel, Johannes Stökl, Rebecca Schweizer, Heiko Vogel, Manfred Ayasse, Joachim Ruther, and Sandra Steiger, "A Hormone-Related Female Anti-Aphrodisiac Signals Temporary Infertility and Causes Sexual Abstinence to Synchronize Parental Care," *Nature Communications* 7 (2016).

13. David Buss, "Sex Differences in Human Mate Preferences: Evolutionary Hypotheses Tested in 37 Cultures," *Behavioral and Brain Sciences* 12 (1989): 1–49.

14. Douglas T. Kenrick, Edward K. Sadalla, Gary Groth, and Melanie R. Trost, "Evolution, Traits, and the Stages of Human Courtship: Qualifying the Parental Investment Model," *Journal of Personality* 58, no. 1 (1990): 97–116.

15. Martin Daly and Margo Wilson, *Homicide* (New Brunswick, NJ: Transaction Publishers, 1988).

16. Heidi Greiling and David M. Buss, "Women's Sexual Strategies: The Hidden Dimension of Extra-Pair Mating," *Personality and Individual Differences* 28, no. 5 (2000): 929–963.

17. Ibid.

18. Kermyt G. Anderson, "How Well Does Paternity Confidence Match Actual Paternity? Evidence from Worldwide Nonpaternity Rates," *Current Anthropology* 47, no. 3 (June 2006): 513–520.

19. Brooke A. Scelza, "Female Choice and Extra-Pair Paternity in a Traditional Human Population," *Biology Letters* (2011): rsbl20110478.

20. Simon C. Griffith, Ian P. F. Owens, and Katherine A. Thuman, "Extra Pair Paternity in Birds: A Review of Interspecific Variation and Adaptive Function," *Molecular Ecology* 11, no. 11 (2002): 2195–2212.

21. Paul W. Andrews, Steven W. Gangestad, Geoffrey F. Miller, Martie G. Haselton, Randy R. Thornhill, and Michael C. Neale, "Sex Differences in Detecting Sexual Infidelity: Results of a Maximum Likelihood Method for Analyzing the Sensitivity of Sex Differences to Underreporting," *Human Nature* 19 (2008): 347–373.

Chapter 5: Mate Shopping

1. Amanda Chan, "How Soay Sheep Survive on Dreary Scottish Isles," *Live Science*, October 28, 2010, https://www.livescience.com/8862-soay-sheep-survive-dreary-scottish-isles.html.

2. Alexandra Brewis and Mary Meyer, "Demographic Evidence That Human Ovulation Is Undetectable (at Least in Pair Bonds)," *Current Anthropology* 46 (2005): 465–471.

3. Daniel M. T. Fessler, "No Time to Eat: An Adaptationist Account of Periovulatory Behavioral Changes," *Quarterly Review of Biology* 78, no. 1 (2003): 3–21.

4. James R. Roney and Zachary L. Simmons, "Ovarian Hormone Fluctuations Predict Within-Cycle Shifts in Women's Food Intake," *Hormones and Behavior* 90 (2017): 8–14.

5. Ibid.

6. Beverly I. Strassmann, "The Evolution of Endometrial Cycles and Menstruation," *Quarterly Review of Biology* 71, no. 2 (1996): 181–220.

7. Andrea Elizabeth Jane Miller, J. D. MacDougall, M. A. Tarnopolsky, and D. G. Sale, "Gender Differences in Strength and Muscle Fiber Characteristics," *European Journal of Applied Physiology and Occupational Physiology* 66, no. 3 (1993): 254–262.

8. Coren Apicella, Elif Ece Demiral, and Johanna Mollerstrom, "No Gender Difference in Willingness to Compete When Competing against Self" (DIW Berlin Discussion Paper 1638, 2017), https://ssrn.com/abstract=2914220.

9. Maryanne L. Fisher, "Female Intrasexual Competition Decreases Female Facial Attractiveness," *Proceedings of the Royal Society B: Biological Sciences* 271, suppl. 5 (2004): S283–S285.

10. Martie G. Haselton, Mina Mortezaie, Elizabeth G. Pillsworth, April Bleske-Rechek, and David A. Frederick, "Ovulatory Shifts in Human Female Ornamentation: Near Ovulation, Women Dress to Impress," *Hormones and Behavior* 51, no. 1 (2007): 40–45.

11. Kristina M. Durante, Norman P. Li, and Martie G. Haselton, "Changes in Women's Choice of Dress across the Ovulatory Cycle: Naturalistic and Laboratory Task-Based Evidence," *Personality and Social Psychology Bulletin* 34, no. 11 (2008): 1451–1460, doi: 10.1177/0146167208323103.

12. Stephanie M. Cantú, Jeffry A. Simpson, Vladas Griskevicius, Yanna J. Weisberg, Kristina M. Durante, and Daniel J. Beal, "Fertile and Selectively Flirty: Women's Behavior toward Men Changes across the Ovulatory Cycle," *Psychological Science* 25, no. 2 (2014): 431–438.

13. Valentina Piccoli, Francesco Foroni, and Andrea Carnaghi, "Comparing Group Dehumanization and Intra-Sexual Competition among Normally Ovulating Women and Hormonal Contraceptive Users," *Personality and Social Psychology Bulletin* 39, no. 12 (2013): 1600–1609.

14. Adar B. Eisenbruch and James R. Roney, "Conception Risk and the Ultimatum Game: When Fertility Is High, Women Demand More," *Personality and Individual Differences* 98 (2016): 272–274.

15. Margery Lucas and Elissa Koff, "How Conception Risk Affects Competition and Cooperation with Attractive Women and Men," *Evolution and Human Behavior* 34, no. 1 (2013): 16–22.

16. Dow Chang, "Comparison of Crash Fatalities by Sex and Age Group," National Highway Traffic Safety Administration, July 2008, https://crashstats.nhtsa.dot.gov/Api/Public/ViewPublication/810853.

17. Diana Fleischman, Carolyn Perilloux, and David Buss, "Women's Avoidance of Sexual Assault across the Menstrual Cycle" (unpublished manuscript, 2017, University of Portsmouth, UK).

18. Sandra M. Petralia and Gordon G. Gallup, "Effects of a Sexual Assault Scenario on Handgrip Strength across the Menstrual Cycle," *Evolution and Human Behavior* 23, no. 1 (2002): 3–10.

19. Daniel M. T. Fessler, Colin Holbrook, and Diana Santos Fleischman, "Assets at Risk: Menstrual Cycle Variation in the Envisioned Formidability of a Potential Sexual Assailant Reveals a Component of Threat Assessment," *Adaptive Human Behavior and Physiology* 1, no. 3 (2015): 270–290.

20. Debra Lieberman, Elizabeth G. Pillsworth, and Martie G. Haselton, "Kin Affiliation across the Ovulatory Cycle: Females Avoid Fathers When Fertile," *Psychological Science* 22, no. 1 (2011): 13–18.

21. Debra Lieberman, John Tooby, and Leda Cosmides, "Does Morality Have a Biological Basis? An Empirical Test of the Factors Governing Moral Sentiments Relating to Incest," *Proceedings of the Royal Society B: Biological Sciences* 270, no. 1517 (2003): 819–826.

22. J. Boudesseul, K. A. Gildersleeve, M. G. Haselton, and L. Bègue, "Do Women Expose Themselves to More Health-Related Risks in Certain Phases of the Menstrual Cycle? A Meta-Analytic Review" (in preparation, 2017).

Chapter 6: The (Not Quite) Undercover Ovulator

1. Alec T. Beall and Jessica L. Tracy, "Women Are More Likely to Wear Red or Pink at Peak Fertility," *Psychological Science* 24, no. 9 (2013): 1837–1841; Pavol Prokop and Martin Hromada, "Women Use Red in Order to Attract Mates," *Ethology* 119, no. 7 (2013): 605–613.

2. Richard L. Doty, M. Ford, George Preti, and G. R. Huggins, "Changes in the Intensity and Pleasantness of Human Vaginal Odors During the Menstrual Cycle," *Science* 190 (1975): 1316–1318.

3. Kelly A. Gildersleeve, Martie G. Haselton, Christina M. Larson, and Elizabeth G. Pillsworth, "Body Odor Attractiveness as a Cue of Impending Ovulation in Women: Evidence from a Study Using Hormone-Confirmed Ovulation," *Hormones and Behavior* 61, no. 2 (2012): 157–166.

4. Steven W. Gangestad, Randy Thornhill, and Christine E. Garver, "Changes in Women's Sexual Interests and Their Partner's Mate-Retention Tactics across the Menstrual Cycle: Evidence for Shifting Conflicts of Interest," *Proceedings of the Royal Society B: Biological Sciences* 269, no. 1494 (2002): 975–982; Martie G. Haselton and Steven W. Gangestad, "Conditional Expression of Women's Desires and Men's Mate Guarding across the Ovulatory Cycle," *Hormones and Behavior* 49, no. 4 (2006): 509–518.

5. Melissa R. Fales, Kelly A. Gildersleeve, and Martie G. Haselton, "Exposure to Perceived Male Rivals Raises Men's Testosterone on Fertile Relative to Nonfertile Days of Their Partner's Ovulatory Cycle," *Hormones and Behavior* 65, no. 5 (2014): 454–460.

6. Martie G. Haselton and Kelly Gildersleeve, "Can Men Detect Ovulation?" *Current Directions in Psychological Science* 20, no. 2 (2011): 87–92.

7. Christopher W. Kuzawa, Alexander V. Georgiev, Thomas W. McDade, Sonny Agustin Bechayda, and Lee T. Gettler, "Is There a Testosterone Awakening Response in Humans?" *Adaptive Human Behavior and Physiology* 2, no. 2 (2016): 166–183.

8. Ana Lilia Cerda-Molina, Leonor Hernández-López, E. Claudio, Roberto Chavira-Ramírez, and Ricardo Mondragón-Ceballos, "Changes in Men's Salivary Testosterone and Cortisol Levels, and in Sexual Desire after Smelling Female Axillary and Vulvar Scents," *Frontiers in Endocrinology* 4 (2013): 159, doi: 10.3389/fendo.2013.00159.

9. Ibid.

10. Kelly A. Gildersleeve, Melissa R. Fales, and Martie G. Haselton, "Women's Evaluations of Other Women's Natural Body Odor Depend on Target's Fertility Status," *Evolution and Human Behavior* 38, no. 2 (2017): 155–163.

11. You can listen to an elephant's estrous rumble here. (Be careful, however: The low frequency could hurt your ears if you wear headphones to listen.) "Estrous-Rumble," Elephant Voices, https://www.elephantvoices.org/multimedia-resources/elephant-calls-database-contexts/230-sexual/female-choice/estrous-rumble.html?layout=callscontext.

12. Gregory A. Bryant and Martie G. Haselton, "Vocal Cues of Ovulation in Human Females," *Biology Letters* 5, no. 1 (2009): 12–15.

13. Nathan R. Pipitone and Gordon G. Gallup, "Women's Voice Attractiveness Varies across the Menstrual Cycle," *Evolution and Human Behavior* 29, no. 4 (2008): 268–274; David A. Puts, Drew H. Bailey, Rodrigo A. Cárdenas, Robert P. Burriss, Lisa L. M. Welling, John R. Wheatley, and Khytam Dawood, "Women's Attractiveness Changes with Estradiol and Progesterone across the Ovulatory Cycle," *Hormones and Behavior* 63, no. 1 (2013): 13–19.

14. C. D. Buesching, M. Heistermann, J. K. Hodges, and Elke Zimmermann, "Multimodal Oestrus Advertisement in a Small Nocturnal Prosimian, *Microcebus murinus*," *Folia Primatologica* 69, suppl. 1 (1998): 295–308.

15. Alan F. Dixson, *Primate Sexuality: Comparative Studies of the Prosimians, Monkeys, Apes, and Humans*, 2nd ed. (New York: Oxford University Press, 2012), 142.

16. Remco Kort, Martien Caspers, Astrid van de Graaf, Wim van Egmond, Bart Keijser, and Guus Roeselers, "Shaping the Oral Microbiota through Intimate Kissing," *Microbiome* 2, no. 1 (2014): 41.

17. Claus Wedekind, Thomas Seebeck, Florence Bettens, and Alexander J. Paepke, "MHC-Dependent Mate Preferences in Humans," *Proceedings of the Royal Society B: Biological Sciences* 260, no. 1359 (1995).

18. Kort et al., "Shaping the Oral Microbiota."

19. Beverly I. Strassmann, "Sexual Selection, Paternal Care, and Concealed Ovulation in Humans," *Ethology and Sociobiology* 2 (1981): 31–40.

20. Joseph Henrich, Robert Boyd, and Peter J. Richerson, "The Puzzle of Monogamous Marriage," *Philosophic Transactions of the Royal Society B* 367, no. 1589 (2012): 657–669.

Chapter 7: Maidens to Matriarchs

1. T. J. Mathews and Brady E. Hamilton, "Mean Age of Mothers Is on the Rise: United States, 2000–2014," *NCHS Data Brief* 232 (2016): 1–8.

2. "About Teen Pregnancy," Centers for Disease Control and Prevention, https://www.cdc.gov/teenpregnancy/about/.

3. Bernard D. Roitberg, Marc Mangel, Robert G. Lalonde, Carol A. Roitberg, Jacques J. M. van Alphen, and Louise Vet, "Seasonal Dynamic Shifts in Patch Exploitation by Parasitic Wasps," *Behavioral Ecology* 3, no. 2 (1992): 156–165, https://doi.org/10.1093/beheco/3.2.156.

4. Bruce J. Ellis, "Timing of Pubertal Maturation in Girls: An Integrated Life History Approach," *Psychological Bulletin* 130, no. 6 (2004): 920.

5. Shannen L. Robson and Bernard Wood, "Hominin Life History: Reconstruction and Evolution," *Journal of Anatomy* 212, no. 4 (2008): 394–425.

6. Lee Alan Dugatkin and Jean-Guy J. Godin, "Reversal of Female Mate Choice by Copying in the Guppy (*Poecilia reticulata*)," *Proceedings of the Royal Society B: Biological Sciences* 249, no. 1325 (1992): 179–184.

7. Jean M. Twenge, *The Impatient Woman's Guide to Getting Pregnant* (New York: Simon and Schuster, 2012).

8. Daniel M. T. Fessler, Serena J. Eng, and C. David Navarrete, "Elevated Disgust Sensitivity in the First Trimester of Pregnancy: Evidence Supporting the Compensatory Prophylaxis Hypothesis," *Evolution and Human Behavior* 26, no. 4 (2005): 344–351.

9. Noel M. Lee and Sumona Saha, "Nausea and Vomiting of Pregnancy," *Gastroenterology Clinics of North America* 40, no. 2 (2011): 309–334.

10. Laura M. Glynn, "Increasing Parity Is Associated with Cumulative Effects on Memory," *Journal of Women's Health* 21, no. 10 (2012): 1038–1045.

11. Elseline Hoekzema, Erika Barba-Müller, Cristina Pozzobon, Marisol Picado, Florencio Lucco, David García-García, Juan Carlos Soliva, et al., "Pregnancy Leads to Long-Lasting Changes in Human Brain Structure," *Nature Neuroscience* 20, no. 2 (2017): 287–296.

12. Chandler R. Marrs, Douglas P. Ferarro, Chad L. Cross, and Janice McMurray, "Understanding Maternal Cognitive Changes: Associations between Hormones and Memory," *Hormones Matter*, March 2013, 1–13.

13. Marla V. Anderson and M. D. Rutherford, "Evidence of a Nesting Psychology During Human Pregnancy," *Evolution and Human Behavior* 34, no. 6 (2013): 390–397.

14. Marla V. Anderson and M. D. Rutherford, "Recognition of Novel Faces after Single Exposure Is Enhanced during Pregnancy," *Evolutionary Psychology* 9, no. 1 (2011), https://doi.org/10.1177/147470491100900107.

15. Jennifer Hahn-Holbrook, Julianne Holt-Lunstad, Colin Holbrook, Sarah M. Coyne, and E. Thomas Lawson, "Maternal Defense: Breast Feeding Increases Aggression by Reducing Stress," *Psychological Science* 22, no. 10 (2011): 1288–1295.

16. Jennifer Hahn-Holbrook, Colin Holbrook, and Martie Haselton, "Parental Precaution: Adaptive Ends and Neurobiological Means," *Neuroscience and Biobehavioral Reviews* 35 (2011): 1052–1066.

17. John G. Neuhoff, Grace R. Hamilton, Amanda L. Gittleson, and Adolfo Mejia, "Babies in Traffic: Infant Vocalizations and Listener Sex

Modulate Auditory Motion Perception," *Journal of Experimental Psychology: Human Perception and Performance* 40, no. 2 (2014): 775.

18. Daniel M. T. Fessler, Colin Holbrook, Jeremy S. Pollack, and Jennifer Hahn-Holbrook, "Stranger Danger: Parenthood Increases the Envisioned Bodily Formidability of Menacing Men," *Evolution and Human Behavior* 35, no. 2 (2014): 109–117.

19. Judith A. Easton, Jaime C. Confer, Cari D. Goetz, and David M. Buss, "Reproduction Expediting: Sexual Motivations, Fantasies, and the Ticking Biological Clock," *Personality and Individual Differences* 49, no. 5 (2010): 516–520.

20. Sindya N. Bhanoo, "Life Span of Early Man Same as Neanderthal," *New York Times*, January 10, 2011, http://www.nytimes.com/2011/01/11/science/11obneanderthal.html.

21. Robson and Wood, "Hominin Life History."

22. Darren P. Croft, Rufus A. Johnstone, Samuel Ellis, Stuart Nattrass, Daniel W. Franks, Lauren J. N. Brent, Sonia Mazzi, Kenneth C. Balcomb, John K. B. Ford, and Michael A. Cant, "Reproductive Conflict and the Evolution of Menopause in Killer Whales," *Current Biology* 27, no. 2 (2017): 298–304.

23. Robin W. Baird and Hal Whitehead, "Social Organization of Mammal-Eating Killer Whales: Group Stability and Dispersal Patterns," *Canadian Journal of Zoology* 78, no. 12 (2000): 2096–2105; Darren P. Croft, Rufus A. Johnstone, Samuel Ellis, Stuart Nattrass, Daniel W. Franks, Lauren J.N. Brent, Sonia Mazzi, Kenneth C. Balcomb, John K.B. Ford, Michael A. Cant, "Reproductive Conflict and the Evolution of Menopause in Killer Whales," *Current Biology* 27, no. 2 (2017): 298–304.

24. Emma A. Foster, Daniel W. Franks, Sonia Mazzi, Safi K. Darden, Ken C. Balcomb, John K. B. Ford, and Darren P. Croft, "Adaptive Prolonged Postreproductive Life Span in Killer Whales," *Science* 337, no. 6100 (2012): 1313.

25. Robson and Wood, "Hominin Life History."

26. Kristen Hawkes and James E. Coxworth, "Grandmothers and the Evolution of Human Longevity: A Review of Findings and Future Directions," *Evolutionary Anthropology: Issues, News, and Reviews* 22, no. 6 (2013): 294–302.

27. R. Sprengelmeyer, David I. Perrett, E. C. Fagan, R. E. Cornwell, J. S. Lobmaier, A. Sprengelmeyer, H. B. M. Aasheim, et al., "The Cutest Little Baby Face: A Hormonal Link to Sensitivity to Cuteness in Infant Faces," *Psychological Science* 20, no. 2 (2009): 149–154.

Chapter 8: Hormonal Intelligence

1. Incidentally, you may wonder why women still have periods on the pill, even though menstruation serves no biological purpose. In fact, some women use the pill to stop their periods — by taking it without the built-in seven-day break. John Rock, a cocreator of the original birth control pill, was a devout Catholic; in order to avoid running up against the Catholic Church, which had banned "artificial" means of birth control and promoted the rhythm method — sex on "safe" days only — he left in the "natural" phase of menstruation. In 1958 the church allowed the pill to be prescribed as a way to treat painful and difficult periods, as it did — and still does — help relieve severe symptoms of menstruation. The church would ban it outright in 1968. Not surprisingly, for a good ten years many Catholic women told their doctors they had painful and difficult periods. Malcolm Gladwell, "John Rock's Error," *The New Yorker*, March 13, 2000, 52.

2. Alexandra Alvergne and Virpi Lummaa, "Does the Contraceptive Pill Alter Mate Choice in Humans?" *Trends in Ecology and Evolution* 25, no. 3 (2010): 171–179.

3. Ibid.

4. Chris Ryan, "How the Pill Could Ruin Your Life," *Psychology Today*, May 11, 2010, https://www.psychologytoday.com/blog/sex-dawn/201005/how-the-pill-could-ruin-your-life.

5. Christina Marie Larson, "Do Hormonal Contraceptives Alter Mate Choice and Relationship Functioning in Humans?" (PhD diss., University of California, Los Angeles, 2014).

6. Shimon Saphire-Bernstein, Christina M. Larson, Kelly A. Gildersleeve, Melissa R. Fales, Elizabeth G. Pillsworth, and Martie G. Haselton, "Genetic Compatibility in Long-Term Intimate Relationships: Partner Similarity at Major Histocompatibility Complex (MHC) Genes May Reduce In-Pair Attraction," *Evolution and Human Behavior* 38, no. 2 (2017): 190–196.

7. Larson, "Do Hormonal Contraceptives Alter Mate Choice?"; Shimon Saphire-Bernstein, Christina M. Larson, Elizabeth G. Pillsworth, Steven W. Gangestad, Gian Gonzaga, Heather Strekarian, Christine E. Garver-Apgar, and Martie G. Haselton, "An Investigation of MHC-Based Mate Choice among Women Who Do versus Do Not Use Hormonal Contraception" (unpublished manuscript). Saphire-Bernstein et al., "Genetic Compatibility in Long-Term Intimate Relationships."

8. Michelle Russell, V. James K. McNulty, Levi R. Baker, and Andrea L. Meltzer, "The Association between Discontinuing Hormonal Contraceptives and Wives' Marital Satisfaction Depends on Husbands' Facial Attractiveness,"

Proceedings of the National Academy of Sciences 111, no. 48 (2014): 17081–17086.

9. Trond Viggo Grøntvedt, Nicholas M. Grebe, Leif Edward Ottesen Kennair, and Steven W. Gangestad, "Estrogenic and Progestogenic Effects of Hormonal Contraceptives in Relation to Sexual Behavior: Insights into Extended Sexuality," *Evolution and Human Behavior* 31, no. 3 (2017): 283–292.

10. Geoffrey Miller, Joshua M. Tybur, and Brent D. Jordan, "Ovulatory Cycle Effects on Tip Earnings by Lap Dancers: Economic Evidence for Human Estrus?" *Evolution and Human Behavior* 28, no. 6 (2007): 375–381.

11. Shannen L. Robson and Bernard Wood, "Hominin Life History: Reconstruction and Evolution," *Journal of Anatomy* 212, no. 4 (2008): 394–425.

12. "Depression among Women," Centers for Disease Control and Prevention, https://www.cdc.gov/reproductivehealth/depression/index.htm.

13. Jennifer Hahn-Holbrook and Martie Haselton, "Is Postpartum Depression a Disease of Modern Civilization?" *Current Directions in Psychological Science* 23, no. 6 (2014): 395–400.

14. Natasha Singer and Duff Wilson, "Menopause, As Brought to You by Big Pharma," *New York Times*, December 12, 2009, http://www.nytimes.com/2009/12/13/business/13drug.html?mcubz=0.

15. Kathryn S. Huss, "Feminine Forever," book review, *Journal of the American Medication Association* 197, no. 2 (July 11, 1966).

16. Joe Neel, "The Marketing of Menopause," NPR, August 8, 2002, http://www.npr.org/news/specials/hrt/.

17. Roger A. Lobo, James H. Pickar, John C. Stevenson, Wendy J. Mack, and Howard N. Hodis, "Back to the Future: Hormone Replacement Therapy as Part of a Prevention Strategy for Women at the Onset of Menopause," *Atherosclerosis* 254 (2016): 282–290.

18. JoAnn E. Manson and Andrew M. Kaunitz, "Menopause Management—Getting Clinical Care Back on Track," *New England Journal of Medicine* 374, no. 9 (2016): 803–806.

19. Robert Bazell, "The Cruel Irony of Trying to Be Feminine Forever," NBC News, 2013, http://www.nbcnews.com/id/16397237/ns/health-second_opinion/t/cruel-irony-trying-be-feminine-forever/#.WUK9cemQyUk.

20. Marcia E. Herman-Giddens, Eric J. Slora, Richard C. Wasserman, Carlos J. Bourdony, Manju V. Bhapkar, Gary G. Koch, and Cynthia M. Hasemeier, "Secondary Sexual Characteristics and Menses in Young Girls Seen in Office Practice: A Study from the Pediatric Research in Office Settings Network," *Pediatrics* 99, no. 4 (1997): 505–512.

21. Louise Greenspan and Julianna Deardorff, *The New Puberty: How to Navigate Early Development in Today's Girls* (New York: Rodale, 2014); Dina Fine Maron, "Early Puberty — Causes and Effects," *Scientific American*, May 1, 2015, https://www.scientificamerican.com/article/early-puberty-causes-and-effects/.

22. Frank M. Biro, Maida P. Galvez, Louise C. Greenspan, Paul A. Succop, Nita Vangeepuram, Susan M. Pinney, Susan Teitelbaum, Gayle C. Windham, Lawrence H. Kushi, and Mary S. Wolff, "Pubertal Assessment Method and Baseline Characteristics in a Mixed Longitudinal Study of Girls," *Pediatrics* 126, no. 3 (2010): e583–e590.

23. Yichang Chen, Le Shu, Zhiqun Qiu, Dong Yeon Lee, Sara J. Settle, Shane Que Hee, Donatello Telesca, Xia Yang, and Patrick Allard, "Exposure to the BPA-Substitute Bisphenol S Causes Unique Alterations of Germline Function," *PLoS Genetics* 12, no. 7 (2016): e1006223; Wenhui Qiu, Yali Zhao, Ming Yang, Matthew Farajzadeh, Chenyuan Pan, and Nancy L. Wayne, "Actions of Bisphenol A and Bisphenol S on the Reproductive Neuroendocrine System during Early Development in Zebrafish," *Endocrinology* 157, no. 2 (2015): 636–647.

24. Paul B. Kaplowitz, "Link between Body Fat and the Timing of Puberty," *Pediatrics* 121, suppl. 3 (2008): S208–S217.

25. Eric Vilain and J. Michael Bailey, "What Should You Do if Your Son Says He's a Girl?" *Los Angeles Times*, May 21, 2015, http://www.latimes.com/opinion/op-ed/la-oe-vilain-transgender-parents-20150521-story.html.

Index

Index

About the Author

Martie Haselton, PhD, is a pioneering researcher on how ovulatory cycles have influenced women's sexuality and how women's hormone changes across the life span affect their social relationships. She is professor of psychology and communication and a professor in the Institute for Society and Genetics at UCLA. She was editor of the journal *Evolution and Human Behavior.* She lives in Los Angeles with her two children, Georgia and Lachlan, fifteen-year-old golden retriever, Biscuit, and much younger tabby cat, Jones.